Basic and Clinical Aspects of Interferon Gamma

Edited by Hridayesh Prakash

Published in London, United Kingdom

Basic and Clinical Aspects of Interferon Gamma
http://dx.doi.org/10.5772/intechopen.98072
Edited by Hridayesh Prakash

Contributors
Filex Odhiambo Otieno, Cynthia Wayua Kyalo, Asimul Islam, Seemi Farhat Basir, Ayesha Aiman, Asesh Banerjee, Prabuddha Gupta, Pragya Santra, Prama Ghosh, Soumyadeep Ghosh, Anwesha Behera, Oyendrilla Mitra, Ishanee Das Sharma, Diya Adhikary, Svetlana Berns, Nadezhda S Astanina, Goar Balasaniantc, Denis V Barbinov, Julija V Dolgo-Saburova, Feliks I Ershov, Natalia A Falaleeva, Ekaterina V Frolova, Alisa I Gil', Lyudmila Yuryevna Grivtsova, Andrej V Ignatovskij, Valentina G Isaeva, Anastasiya A Kalashnikova, Vera V Kir'Janova, Tat'Jana B Korzhenevskaja, Irina A. Rakityanskaya, Tatiana S. Ryabova, Elena V Shagdileeva, Natal'Ja V Vasil'Eva, Kaprin A.D, Vijay Kumar, Hridayesh Prakash, Approv Sharma, Abhishek K. Singh

First published in London, United Kingdom, 2022 by IntechOpen
IntechOpen is the global imprint of INTECHOPEN LIMITED, registered in England and Wales, registration number: 11086078, 5 Princes Gate Court, London, SW7 2QJ, United Kingdom

British Library Cataloguing-in-Publication Data
A catalogue record for this book is available from the British Library

Additional hard and PDF copies can be obtained from orders@intechopen.com

Basic and Clinical Aspects of Interferon Gamma
Edited by Hridayesh Prakash
p. cm.
Print ISBN 978-1-80355-885-1
Online ISBN 978-1-80355-886-8
eBook (PDF) ISBN 978-1-80355-887-5

Meet the editor

Prof. Hridayesh Prakash is a fellow of the Royal Society of Biology, London. Currently, he is a professor in the Department of Biotechnology, MM College of Engineering, India. He has expertise in innate immunity with a special interest in macrophage immunobiology, tumor immunology/immunotherapy, cell-based immunotherapies, pulmonary infection biology, and radiation biology. Prof. Prakash conducts research to exploit various immunotherapeutics for managing persistent bacterial and viral infections and gastric cancer. He is unraveling the therapeutic potential of M1 effector macrophages against solid tumors. He is also studying various mechanisms that certain pathogens like *Helicobacter pylori*, chlamydia, and mycobacteria are exploiting for polarizing M1 effector macrophages towards the M2 phenotype during chronic and persistent infections. Under this major objective, he is now validating the therapeutic impact of M1 effector macrophages for the control of persistent infection-driven cancer (adenocarcinoma) progression. Prof. Prakash is also exploring the palliative potential of macrophages against autoimmunity and chronic inflammatory disorders like inflammatory bowel disease, radio-pneumonitis, pulmonary fibrosis, and radiation syndrome.

Contents

Preface

The immune system is one of the plastic systems of the body and is heavily engaged in managing equilibrium in all systems. Interferon (IFN) is the key soluble component of the immune system, a potent inducer of tumor necrosis factor (TNF) and the interleukin network, and, by and large, sufficient to control various infections, cancer, and immune deficiencies.

The double-edged IFN signaling network largely dictates the plasticity in the immune system by tuning the system's functioning to orchestrate immunity against invaders and cancers while maintaining homeostasis. Apart from managing infections and cancer, the IFN system has the potential for managing a large variety of infectious diseases including COVID-19, in which type I IFN/gamma IFN rheostat can limit the spread of the infection in the host while adjusting the host immunity in its favor.

Several recent studies have demonstrated the impact of IFN signaling on immune-related metabolism, for which underlying molecular/immune mechanisms are not clear. As such, this book discusses IFN biology, chemistry, and physiological aspects, all of which are important for immune-related metabolism in the host. The book also emphasizes how IFN can assist various targeted interventions, harness their therapeutic potential, and benefit the health sector. It examines the plausible impact of GAS(an Interferon gamma-activated site) and/or IRES(Internal Ribosome Entry Sites) elements with miRNA/lncRNA components for controlling the immune-related metabolism of host cells.

Hridayesh Prakash
Department of Biotechnology,
MM College of Engineering,
Maharishi Markandeshwar University,
Mullana, Ambala, Haryana, India

Section 1

Interferon - Genesis

Chapter 1

Perspective Chapter: Macrophages Plasticity and Immune Metabolism

Filex Otieno and Cynthia Kyalo

Abstract

Macrophages are phagocytic cells that reside within body tissues. They can either be derived from circulating monocytes or can arise during the embryonic stage of fetal development. Tissue macrophages are predominantly of embryonic origin. But can result from differentiation of circulating monocytes to become resident macrophages either in pathological or physiological state. Macrophages are classified based on their tissue location and method of activation. Classically activated macrophages are the M1 phenotype while alternatively activated macrophages are M2 phenotype. M1 macrophages are pro-inflammatory since they secrete cytokines that attract inflammatory mediators. They are majorly activated by either interferon-gamma or lipopolysaccharide molecules. M2 macrophages are anti-inflammatory and mediate tissue healing and repair. They are activated by cytokines such as interleukin four, ten, and thirteen. The metabolic profiles of these classes of macrophages are intrinsically different and complex yet intertwined. M1 macrophages depend on aerobic glycolysis for energy production while M2 macrophages rely on aerobic fatty acid oxidation pathways. These metabolic pathways optimize macrophage functioning. Regulation of both activation and metabolism depends on transcriptional factors such as STAT 1 and 6, and IRF. Defects in these pathways lead to development of disorders related to macrophage activation and metabolism.

Keywords: polarization, glycolysis, cytokines, immune metabolism, Krebs cycle, activation, inflammation, diapedesis, translocation, margination, inflammation, immune surveillance

1. Introduction

The immune system is responsible for protection against infections and diseases. It is made up of a complex network of cells and proteins, working in harmony to protect the body against pathogens. White blood cells make up the larger component of this system comprising granulocytes and agranulocytes. Human macrophages can be formed through the differentiation of monocytes. Monocytes are a population of mononuclear leukocytes that are generated in the bone marrow. They constitute approximately 10% of peripheral blood cells in the human body. The monocytes move into the blood from the bone marrow, where they migrate to various body tissues through blood circulation. Once in the tissues, they differentiate into different types of macrophages depending on resident tissue. Some tissue-resident macrophages are non- monocyte-derived. However, their origin, proliferation, self-renewal, and mechanisms of replacement are vaguely well known and will be elaborated on later on in this chapter.

The term "macrophage" is a combination of two Greek words: *makros* which means "large" and *phagein* which means "to eat". They are thus cells that carry out phagocytosis through engulfment and digestion of any pathogen or foreign particle that is opsonized. These include cellular debris, foreign substances, cancer cells, and microbes. Ilya Metchnikoff, a Russian Zoologist was the first to discover macrophages in the nineteenth century (1884) [1]. Van Furth in the 60s proposed that tissue macrophages were effector cells derived from circulating monocytes [1]. Evolutionary, macrophages are conserved phagocytes that have existed for the last 500 million years and belong to the Metazoan phylogeny. Studies in recent years have shown that the majority of resident-tissue macrophages are established at the embryonic stage and persist into adulthood [1]. The function of blood monocytes that get differentiated into macrophages is to replenish the lost pool or aid in situations such as inflammation.

2. Formation and development of macrophages

Macrophages form the major part of the mononuclear phagocytic system. These cells are distributed widely within the body in various organs displaying heterogeneity in terms of their structure and function. Their sizes range between 15 to 50 microns with an ovoid nucleus of about 6 to 12 microns [2]. The cytoplasm is granulated with vacuoles located towards the periphery of the cell. The granules are finely divided especially towards the periphery of the cytoplasm and azurophilic in nature [1]. Macrophages can either arise from the differentiation of recruited macrophages or early in life through macrophage progenitor cells of embryonic origin.

2.1 Blood derived macrophages

Macrophages can be derived from blood monocytes that have migrated from the circulatory system into the tissues. Monocytes are white blood cells formed through hematopoiesis in the bone marrow. They form approximately 10% of the white blood cell composition circulating within the blood, the spleen, and the bone marrow. Structurally, monocytes are irregularly shaped with a kidney-shaped nucleus besides a high ratio of cytoplasm to the nucleus.

Precursor cells for monocytes arise from the bone marrow in a series of steps activated by cytokines and stimulating factors. Sequentially four stages are involved in the formation and development of macrophages i.e. hematopoietic stem cells develop into the common myeloid progenitor cells which then differentiate into granulocyte-macrophage progenitor cells [3]. The latter group of cells differentiates into the common macrophage and dendritic cell precursor before finally committing to the committed monocyte progenitor group. The newly formed monocytes stay within the bone marrow for the first 24 hours before being recruited into circulation [3]. The mature monocyte remains in circulation for about 1 to 2 days before translocating into the tissues. If after that period they have not translocated, the circulating monocytes die and are removed from the blood circulation.

Sequential formation of these cells is governed by cytokines and stimuli factors resulting in either expression of new or loss of specific surface proteins.

Colony-stimulating factor 1 (CSF-1) (also known as monocyte colony-stimulating factor- M-CSF) largely influences the development of monocytes under normal healthy hemostatic conditions. The factor is secreted by bone marrow stromal cells and also in tissues. CSF-1 is removed from circulation by the mononuclear phagocytes reducing hematopoiesis of monocytes. Interleukin 34 (IL-34) found in the central nervous system and the epidermal tissues is responsible for the proliferation of macrophages from monocytes within that specific tissue type. Interleukin 1 (IL-1) and tumor necrosis factor (TNF) on the other hand induce endothelial cells and fibroblast to secrete CSF-1 and CSF-2 [4]. During inflammatory periods, the formation and development of monocytes till their differentiation into macrophages is governed by the granulocyte-macrophage colony-stimulating factor (GM-CSF/CSF-2) and interleukin 3 [4].

Mature monocytes differ in terms of expression of CD16 and CD14 molecules [3, 4]. Classical monocytes express a high CD14 but lack expressed CD 16. They form the major portion of the total monocytes (approx. 90%). Intermediate monocytes lowly express CD16 molecules but high CD14 expression. Alternative monocytes highly express CD16 molecules but CD14 in low amounts. Classical and intermediate monocytes are inflammatory mediators highly expressing the chemokine receptor CCR2 while the alternative monocytes function to patrol the circulatory system for foreign bodies and express the chemokine receptor CX3CR1 [3].

Monocytes that migrate into tissues from the circulatory system differentiate to macrophages. Translocation of monocytes to tissue spaces involves endothelium adherence, diapedesis in between the endothelium, and subsequent migration through the sub-endothelial layer into interstitial space. Monocytes express ligands e.g.CD11a/CD18 (lymphocyte function-associated antigen 1) that bind receptors e.g.CD54 (intercellular adhesion molecule-1) located on the surface of endothelial cells [3]. Interleukin 1 and interferon-gamma (IFN-γ) enhance the expression of CD54 further increasing the margination of monocytes. Upon arrival at the target site, these monocytes differentiate into macrophages. Differentiation largely affects cellular respiration and phagocytic activity. Mitochondria increase in number and their enzymes also show an enhanced rate of activity which ultimately leads to an increased rate of respiration. In addition, lysosomes and their enzymatic activity also increase.

2.2 Macrophages derived from embryonic tissue

Macrophages can also arise from progenitor cells of embryonic origin. These cells are found in almost all tissue types forming approximately 12% of the tissue cells [4]. Depending on the tissues, the macrophages are named differently and epigenetics studies illustrate each of these different tissue macrophages have different transcriptional profiles but all perform similar functions [1, 4]. Their gene expression profile is different from that expressed by bone-marrow-derived macrophages. Primitive hematopoiesis that occurs in the ectoderm of the yolk sac during embryogenesis results in the formation of macrophages that do not follow the monocytic system of development. During subsequent fetal development, definitive hematopoiesis occurs in the fetal liver to give rise to hematopoietic progenitors and stem cells. These cells afterward colonize the spleen and bone marrow taking residency. In adults, the bone marrow becomes the major site of hematopoiesis.

Adult tissue macrophages that were formed from the progenitor cells in the yolk sac and fetal liver have the capacity for self-renewal. CSF-1 and IL-34 still mediate activation and proliferation of these progenitor cells into macrophages depending on the transcription factor PU.1 [1]. These cells upon stimulation directly convert into macrophages. As such, the pool of tissue macrophages is minimally dependent on the translocation of monocytes during steady-state conditions. Macrophages irrespective of origin have to be activated before they can function properly. Macrophages can be activated classically or alternatively.

3. Activation of macrophages

Activation of macrophages comprises an enhanced cellular metabolism, lysosomal activity, mobility, and cytotoxic activity. Activation of macrophages requires a prior recognition of pathogens. Macrophages express an array of receptors that do not normally occur in a healthy cell. These receptors are generally termed pattern recognition receptors (PRRs). They include the toll-like receptors (TLRs), clathrin receptors (CLRs), NLRs, scavenger receptors, and, retinoic acid-inducible gene 1-like helicase receptors. These receptors receive a signal from the damage-associated molecular patterns (DAMPS) and pathogen-associated molecular patterns (PAMPS) present on damaged cells and invading pathogens respectively [5]. PRRs such as mannose receptors are crucial for the process of pathogen binding and phagocytosis. NLRs, TLRs, and RLRs on the other hand are key for the recognition of microbial by-products expressed on other cell surfaces. Though there exists a variety of the PRRs, upon stimulation, the downstream mechanisms result in activation of transcriptional pathways involving mitogen-activated protein kinases, interferon regulatory factors (IRF), and nuclear Factor-kappa B (NF-Kb) [5]. These pathways ultimately lead to the expression and secretion of specific cytokines, phagocytosis, chemokine secretion, cellular activation, and secretion of inflammatory mediators.

Macrophages are activated through two pathways depending on their polarization. Microphage polarization refers to the ability of macrophages to differentiate into functionally different phenotypes [5]. Activation through polarization enables macrophages to optimize the cell to perform a particular function. Macrophages can be polarized to result in the formation of two distinct groups of cells: M1 macrophages and M2 macrophages [6]. The latter is further divided into M2a, b, and c types. M1 macrophages are classically activated while M2 cells are alternatively activated. This classification is based on the metabolic functions of macrophages that can broadly be classified as either pro-inflammatory in nature or anti-inflammatory in nature.

Metabolism of arginine in M1 cells results in the formation of citrulline and nitric oxide. Nitric oxide has a microbicidal activity besides inhibiting cellular proliferation. On the other hand, in M2 macrophages, the metabolism of arginine shifts to result in the formation of polyamines and ornithine. Ornithine enhances cellular proliferation and healing by mediating collagen synthesis, cell repair through polyamines, and fibrosis. By default, resident macrophages are usually polarized to form predominantly M2 macrophages. M1 macrophages are formed in situations where inflammation is warranted such as in cases of infections [7].

3.1 Classical activation

Classical activation of macrophages results in the formation of M1 macrophages which are pro-inflammatory in nature. Ligands such as lipopolysaccharide (LPS) and IFN-γ are key stimulators of macrophages towards the M1 pathway [8]. These ligands stimulate TLRs present on the surface of macrophage membranes resulting in a downstream pathway that leads to the production of chemokines (CXCL2, CCL5, CXCL4, and CCL8), and cytokines (IL-6, TNF-alpha, IL-23, IL-1β, IL-12) [8, 9].

In addition, the pathways activate the expression of inducible nitric oxide synthase (iNOS) leading to the production of nitric oxide. M1 macrophages also express high levels of major histocompatibility molecules II (MHC II) molecules and CD80/86 making them functionally potent antigen-presenting cells (APCs). Six transcriptional factors are activated upon stimulation of TLRs by PAMPS. These are the Signal transducer and activator of transcription 1 (STAT-1), NF-kB, PU.1, IRF5, Activator protein 1 (AP-1), and CCAAT/enhancer-binding protein alpha (C/EBP α) [5, 8, 9]. NF-kB is activated upon stimulation of macrophages by infectious agents, stress, or inflammatory cytokines. Activated NF-kB activates genes that ultimately lead to the relocation and activation of macrophages in the affected region [10] through modulation of the expression of mediators of inflammation and also macrophage differentiation.

When macrophages are exposed to growth factors, infectious agents, cytokines, apoptotic products, and oncogenic stimuli, the transcription factor AP-1 is activated [9]. This factor reduces angiogenesis in a tumor by reducing the production of reactive oxygen species (ROS). The factor also induces the expression of TNF-α during inflammatory responses and also induces mitogen-activated protein kinases (MAPK) thereby stabilizing mRNA that encodes inflammatory cytokines.

PU.1 is a key transcriptional factor during hematopoiesis. It upregulates the expression of the M-CSF receptor thereby enhancing the proliferation of macrophages [3, 5]. In addition, the factor also modulates the expression of genes responsible for the differentiation of macrophages by regulating the expression of GM-CSF receptors thereby affecting the maturation of such cells. In cases where the factor is inappropriately expressed or deficient, carcinogenesis may occur leading to the development of lymphomas.

3.2 Alternative activation

Activation of macrophages in the direction of M2 polarization is predominantly done by cytokines including IL-4, IL-33, IL-13, and IL-10 [3, 5, 8]. M2 macrophages have anti-inflammatory and tissue repair roles within the body. During tissue injury or in response to fungal and parasitic chitin or as a result of adaptive immune responses, IL-4 is released by cells such as basophil and mast cells as one of the earlier mediators of wound healing. This cytokine binds to its receptors on the surface of macrophages inducing a downstream signaling pathway that results in stimulation of arginase activity. Arginase enzyme converts arginine in macrophages to ornithine which is a precursor for the formation of polyamines and collagen. As a result, the increased levels of ornithine are used to form collagen fibers necessary for the wound healing process [7].

In response to parasitic and fungal chitin, the downstream pathway leads to the activation of chitinase enzymes that aid in the interruption of parasitic or fungal cellular integrity. However, Dysregulated activation of macrophages via this pathway can lead to excessive deposition of extracellular matrix leading to fibrosis [7]. Activated M2 macrophages thus secrete IL-1 antagonist, TGF-β, and IL-10 besides blocking iNOS. IL-4 and -13 induce the development of M2a macrophages responsible for wound healing. Immune complexes formed from microbial ligands binding to TLR activate macrophages to develop into the M2b phenotype. However, such cells in presence of IL-10, glucocorticoids, and TGF-βdifferentiate into M2cphenotypes [5].

Five transcriptional factors enhance the activation of macrophages towards the M2 phenotype. These factors include *STAT6, IRF4*, Krüppel-like factor 4 (*KLF4*), *C/EBP-β*, and peroxisome proliferator-activated receptors (*PPAR-γ*) [11]. *KLF4* is part of the zinc family transcription factors and regulates the development, differentiation, and activation of cells. *IRF4* regulates the expression of interferon genes. *IRF4* has a pivotal role during the polarization of macrophages towards the M2 phenotype. The binding of IL-4/IL-13 to their receptors induces phosphorylation and subsequent dimerization of *STAT6* that in turn recruits IRF4 stimulating expression of M2-associated genes such as mannose receptor and chitinase receptor. In addition, the IL4-STAT6 pathway induces expression of nuclear receptor peroxisome proliferator-activated gamma which inhibits the transcriptional factors *AP-1, STAT,* and *NF-kB*. The IL4-STAT6 phosphorylation also promotes *KLF4* expression which leads to the expression of the *M2 gene*.

4. Types of macrophages

Microphages can be classified based on the method of activation or based on the resident tissues where they are located. Though macrophages, in general, have similar functions, they are specialized to perform specific tissue functions [12].

4.1 Classification based on location

Macrophages are given different names depending on the tissue they reside in. **Table 1** below shows the different macrophages, their location, and their functions.

Type of macrophage	Residing body tissue	functions
Microglia	Brains (central nervous system)	Development of brain, Immune surveillance, Remodeling of brain synapses
Osteoclasts	Bone marrow tissue	Remodeling and resorption of bone tissue. Hematopoiesis [12]
Heart macrophages	Cardiovascular system	Immune surveillance
Kupffer cells	Liver	Remove toxins Lipid metabolism Recycling of iron Clearance of microbes, debris, and red blood cells.
Alveolar macrophages	Lungs	Clearance of surfactant, Immune surveillance
Adipose tissue-associated macrophages	Adipose tissue	Lipid metabolism, Adipogenesis, Thermogenesis [13]
Bone marrow macrophages	Bone marrow tissue	Reservoir of monocytes waste disposal [12]
Intestinal macrophages	Gut tissue	Development of microbiota tolerance, Intestinal homeostasis, Defense against pathogens [12]
Langerhans cells	Skin tissue	Immune surveillance
Marginal zone macrophages	Spleen	Trapping of microbes from blood, Red blood cell clearance, Iron processing

Type of macrophage	Residing body tissue	functions
Tumor-associated macrophages(TAM)	Tumors	Create an immunosuppressive tumor microenvironment through the production of cytokines, chemokines, and growth factors
$CD169^+$ macrophages	Lymphoid organs and tissue	involved in immune tolerance antigen presentation immune system regulation erythropoiesis regulation [13]
TCR^+ macrophages		Strong phagocytic activity

Table 1.
Classification of macrophages based on their location.

Type of microphage	Method of activation	Activators
M1 macrophages	Classical activation	IFN-γ from Th_1 cells, cytotoxic T lymphocytes, Nk cells. Lipopolysaccharide [8]
M2 macrophages	Alternative activation	exposure to IL-4, IL-10 and IL-13, CSF-1, TFG-beta Fungal and helminthic infections [11]

Table 2.
Classification of macrophages based on method of activation.

4.2 Classification based on activation

Microphages can be either activated classically or alternatively. The activation depends on the inducing stimuli. In addition, specific cytokines also activate macrophages as shown in the **Table 2**.

5. Functions of macrophages

Macrophage's main role involves maintenance of the integrity of an organism either by direct pathogen elimination or tissue repair under sterile inflammatory conditions.

Macrophages are immune cells tasked with eliminating pathogens within the body and damaged body cells. This they do in various specific ways but in general, will lead to pathogen elimination.

5.1 Immune surveillance

Macrophages recognize PAMPs and DAMPs through their wide array of receptors. As professional phagocytes, they phagocytose pathogenic microorganisms and cellular debris. The process involves phagosome formation through endocytosis. The phagosome is then fused with a lysosome forming phagolysosomes, where the pathogen is enzymatically broken down.

5.2 Induction of inflammation

Macrophages take part in the initiation of inflammatory responses. Once the macrophages have detected possible infectious agents and subsequently bind the PAMPs, they are activated to start producing proinflammatory cytokines. The resident macrophages together with other resident immune cells, such as mast cells, stromal cells, and dendritic cells, initiate an influx of inflammatory leukocytes into the site mainly neutrophils and cytotoxic T lymphocytes.

5.3 Adaptive immunity

Macrophages and dendritic cells are antigen-presenting cells (APCs). Damaged cells and pathogens, once ingested are broken down into smaller molecules. These molecules are complexed with MHC molecules and expressed on the surface of macrophages. The expressed molecules act as antigens that stimulate lymphocytes. Lymphocytes recognize the antigens, become activated, and proliferate to effector cells that eliminate the pathogen.

5.4 Wound healing

Functions of M2 macrophages predominantly involve wound repair and anti-inflammatory action. Monocytes from the bloodstream are recruited to a wounded tissue by growth factors secreted by platelets and other cells at the site and then mature into macrophages. They degrade pathogen or cellular debris available to clear them from the tissue. They also secrete growth factors and cytokines that attract fibroblast involved in healing. The low oxygen concentration of the wounded tissue surroundings stimulates macrophages to secrete cytokines that induce and speed up angiogenesis. They also stimulate granulation of affected tissue by laying down new extra-cellular matrices.

5.5 Tissue homeostasis

Tissue-resident macrophages are specialized both physically and functionally to the role they play in the specific tissues they reside. The resident macrophages are non-migratory residing permanently at the tissue they are adapted to function. They support the physiological function of the tissue by providing essential growth factors. They also actively protect the tissue from inflammatory damage.

5.6 Iron homeostasis

Red blood cells live up to 120 days after which they are destroyed and removed from the circulation in the spleen and liver. The macrophages carrying out this role can engulf macromolecules. They play a role in the control of the distribution of parenteral irons. Iron released from destroyed erythrocytes is stored internally in ferritin or released into circulation via ferroportin. During inflammation, the elevated level of hepcidin leads to the retention of iron in the macrophages. This is through the regulation of macrophage ferroportin channels.

5.7 Muscle regeneration

Phagocytic macrophages are recruited during periods of increased muscle use, causing muscle membrane lysis and membrane inflammation. Their peak concentration at the site is usually reached about 4 hours following the onset of some form of muscle cell injury or reloading. They degrade the contents of injured muscle fibers. M2 macrophages on the other hand are usually distributed near regenerative fibers.

They release soluble substances that influence proliferation, differentiation, growth, repair, and regeneration of muscle [1]. The number of M2 macrophages remains elevated for several days until muscle tissue rebuilding is done.

5.8 Pigment retention

Melanophages which are a subset of macrophages can absorb and retain pigments. The pigment could either be native to the organism or exogenous from extracellular spaces such as tattoo ink. The pigment from dead dermal macrophages is phagocytosed by their successors thereby accumulating phagocytosed melanin in lysosome-like phagosomes. The ultimate effect is the retention of the pigment at the same place [14].

6. Macrophage immune metabolism

Macrophages can be found in nearly all the tissues in the body. The microenvironment around such tissues dictates the metabolic profile of macrophages which in turn affects their function. Moreover, the microenvironment also influences polarization that a microphage cell will undertake within a given tissue. The influence of tissue microenvironment on macrophage functions is indirect through modulation of their metabolic profile. The genetic makeup of macrophages regulating their metabolic pathway is highly plastic due to the influence of the micro-environment [7]. This makes microphages sensitive to metabolic changes both intracellularly and extracellularly. Moreover, incoming monocytes from systemic circulation that are translocating into the tissue are influenced by the niches within the various organs to differentiate into tissue-specific macrophages ensuring the macrophage pool of a given tissue is not exhausted.

Cellular metabolism comprises a network of pathways that utilize fuels from nutrients for energy production and structural formation. Immuno-metabolism illustrates the sub-domain of the immune system that looks at how bioenergetics, usage of nutrients, and metabolite generation all influence and impact the function of immune mediators-in this case, the macrophages. As such, the functional phenotype of a macrophage expressed in a given tissue can be said to be the resultant effect of local environmental signals and metabolic settings. This versatility is attributable to a large number of metabolic and transcriptional profiles. Macrophages, just like any other cell, utilize fuels (glucose, amino acids, and lipids) to generate metabolites (pyruvate, TCA metabolites) and energy in form of ATP through biochemical processes such as glycolysis, and glutaminolysis.

6.1 Intracellular mechanism of macrophage metabolism

Three characteristics are manifested by M1 macrophages activation: secretion of proinflammatory cytokines, expression of iNOS, and cellular metabolism through glycolysis [12, 15]. The major stimulating signals are LPS and IFN-γ. These ligands bind to their receptors and upon ligation, such macrophages undergo increased uptake of glucose [16]. This activates the glycolytic pathway leading to increased formation of glycolytic intermediates. The intermediates are thus shifted into the pentose phosphate pathway (PPP) which leads to the regeneration of NADPH used for nucleotide synthesis and production of ROS. In addition, excess amounts of pyruvate molecules are produced. Some of these molecules are converted into lactate while others enter the Tri-carboxylic acid cycle. However, the pyruvate that enters the TCA cycle is not completely metabolized through the pathway rather two breaks occur within the cycle that shifts into other pathways [12, 17]. One of the breaks occurs during the conversion of isocitrate to alpha-ketoglutarate by the enzyme isocitrate dehydrogenase [12, 15, 18]. Macrophages downregulate this reaction step leading to the accumulation of citrate. The accumulated citrate is in turn shunted to the pathway used to form itaconate which is a potent microbicidal metabolite. The excess citrate is also converted to Acetyl-CoA which is used to acetylate histones within inflammatory genes. This acetylation activates the expression of inflammatory genes. The formed Acetyl-CoA is also used for de novo synthesis of fatty acids that are used to expand cell membranes, synthesize prostaglandins and arachidonic acids.

The second break occurs in the reaction step where succinate is converted into fumarate by succinate dehydrogenase [12]. The earlier formed itaconate mediates this inhibition leading to the accumulation of succinate metabolite. Increased amounts of succinate molecules function as stabilizers for the hypoxia-inducible factor-1α (HIF-1-α) [18]. This in turn decreases oxygen tension within tissues leading to a decrease in mitochondrial respiration. Thereby, macrophages have to be dependent on glycolysis for energy production. The accumulated HIF-1-αpromotes the transcription of IL-1β (a pro-inflammatory cytokine) and enhances the expression of hexokinase1 and glucose transporter1(GLUT1) [18]. This increases the uptake and metabolism of glucose via the glycolytic pathway. Moreover, the little oxidation of succinate by succinate dehydrogenase in macrophages stimulated by LPS induces the generation of mitochondrial reactive oxygen species (mtROS) [18, 19]. These byproducts are recruited into phagosomes for bacterial killing. In addition, the ROS cause oxidative damage to the DNA, and as a result, the poly (ADP-ribose) polymerase (PARP) enzymes are activated [19]. These enzymes consume a lot of NAD leaving the M1 macrophages to rely on salvage pathways for NAD^+. Apart from succinate, the induced nitric oxide also reduces mitochondrial respiration. This leads to a reduction in the ATP: ADP ratio dampening the inflammatory process.

Activation of macrophages towards the M2 phase is often induced by IL-4 and IL-13 [5, 17, 18]. Such cells have enhanced mitochondrial respiration, production of anti-inflammatory cytokines, and increased expression of arginase-1. The increased arginase-1 activity shifts the metabolism of arginine to result in the formation of ornithine metabolites. The ornithine is used in the biosynthesis of various polyamines such as spermidine [18]. Spermidine activates eukaryotic initiation factor 5A (EIF-5A) which is a translation factor. This factor facilitates the expression of mitochondrial proteins needed for the oxidative phosphorylation-dependent differentiation of M2 cells [20]. The overall effect is increased energetic profile of the macrophage and this energy is shifted towards uptake, transport, and oxidation of fatty acids besides increased glucose uptake. M2 macrophages unlike M1 macrophages rely on glutamine and fatty acid oxidation for metabolic processes, especially if the activating stimuli is IL-4.

Transcriptional factors such as *STAT6, IRF-4,* and *PPARγ* are highly activated and expressed in M2 macrophages as earlier discussed [5, 18]. These factors in addition increase the expression of CD36 on surfaces of M2 macrophages. CD36 enhances the endocytosis of lipoproteins containing triglycerides. This increased uptake of lipids leads to an even higher rate of fatty acid oxidation and mitochondrial biogenesis enhancing the consumption rate of oxygen within cells.M2 cells also have a higher rate of glutamine catabolism resulting in the accumulation of alpha-ketoglutarate which is one of the metabolites of the TCA cycle. Alpha-ketoglutarate in turn activates the jumonji domain-containing protein D3 demethylases which promote histone demethylation on M2-specific gene promoters which as a result leads to macrophage polarization towards the M2 direction.

7. Role of IFN-γ in macrophage activation, metabolism, and functioning

Interferons are a group of potent cytokines that are pleiotropic in nature and carry out paracrine and autocrine actions [21]. Their secretion is majorly induced during viral infection but also to a lesser extent during bacterial or autoimmune disorders. Three types of interferons exist IFN α, -β, and -γ. The former two are grouped and known as type I interferons while the latter is commonly grouped under type II interferons [21]. The three molecules exhibit different primary molecular structures. IFN-α, and -β have more or less similar properties, unlike IFN-γ [21]. These different characteristics and properties are illustrated in the **Table 3**.

Type I interferons do not have much of important roles in macrophage activation and metabolism as seen it has on plasmacytoid dendritic cells [21]. IFN-γ on the other hand regulates the activation and metabolic profile of macrophages. IFN-γ does this indirectly by either activating macrophages towards the M1 direction and upregulating or downregulating the expression of genes that enhance energy production via the glycolytic pathway. IFN-γ binds to its receptor IFN-γR activating Janus Kinase 1 and 2 associated protein tyrosine kinase [21, 22]. This causes phosphorylation of the tyrosine residues and activation of signal transducer and activator of transcription-1 (STAT1). STAT1 translocates into the nucleus and activates IFN-y related interferon-stimulated genes. As a result, the activated genes encode for various proteins that manifest the downstream effect of IFN-y on macrophage metabolism.

IFN-γ acts on three major enzymes when it comes to modulating macrophage metabolism [22]. These are the mammalian target of rapamycin complex 1 (mTORC1), glycogen synthase kinase 3 (GSK3), and 5'-AMP-activated protein kinase (AMPK) [23]. IFN-γ inhibits the activity of mTORC1 in macrophages resulting in decreased mitochondrial function, decreased secretion of anti-inflammatory mediators, and synthesis of purine nucleotides. In addition, this inhibition to leads to more autophagy by macrophages enhancing the process of microbial killing. GSK3 upon stimulation by IFN-γ modulates the activity of NF-kB such that an increase in the production of inflammatory cytokines is seen [21]. AMPK acts as a control point. It senses energy deprivation during metabolic processes in M1 macrophages and shifts

IFN α/β	IFN-γ
Their genes are not split by introns	The gene is split by introns
The molecule is stable at pH 2.0	It is acid-labile
Both have a common receptor different from that of IFN-γ (IFNAR1 and IFNAR2)	Its receptor is different from that of type I interferons (IFNGR1 and IFNGR2)
Stimulate mainly natural killer cells	Stimulates manly macrophages
Rapid induction of an antiviral state	Slow induction of an antiviral state
Mainly secreted by IFN-α (B-cells and macrophages), IFN-α(epithelial cells and fibroblasts)	Mainly Secreted by T cells and NK cells

Table 3.
Differences between type 1 interferons and type II interferons.

the metabolism towards the M2 direction [23]. As a result, aerobic respiration and energy production are revamped within the cells.

8. Disorders associated with dysfunctional macrophages

Macrophages just like any other cell can be dysfunctional. These disorders can directly affect the activation, metabolism, and functions of macrophages or indirectly aid in the development of systemic diseases such as metabolic syndromes.

Activation of macrophages leads to their expansion to different functional cell lines. Disordered expansion of macrophages leads to a collective group of disorders known as hemophagocytolymphohistiocytosis (HLH) [24]. This group of illnesses can either be familial/primary or secondary/reactive. Familial HLH is due to inherited autosomal recessive immune disorders caused by genetic defects of genes controlling the cytolytic pathway of macrophages. Secondary HLH is caused by infections particularly Epstein-Barr virus (EBV), Cytomegalovirus (CMV), and cancer [25].

One of the disordered related to macrophage activation is the macrophage activation syndrome [25]. This syndrome is caused by too much activation and subsequent differentiation of macrophages. Clinically it manifests as cytopenias, dysfunctional liver state, hyperferritinemia, and coagulopathy [26]. It is a life-threatening condition with mortality rates of between 20 and 30%. This syndrome occurs mostly and is associated with systemic lupus erythematosus (SLE), systemic juvenile idiopathic arthritis (SJIA), Kawasaki disease, and other rheumatic conditions [27]. Cytotoxic cells induce apoptosis of activated macrophages. Failure of this to occur leads to prolonged expansion of macrophages leading to excessive production of proinflammatory cytokines. This induces a state of cytokine storm and hemophagocytosis is related to changes in metabolic pathways of macrophages [13]. The alveolar macrophages first utilize glycolysis for energy production but then switch to fatty acid metabolism. This process is dependent on the mitochondrial calcium channel and levels of mitochondrial reactive oxygen species. This activation sustains fibrosis within the ling tissues. However, administration of itaconate or when systemized by the macrophages reverses the glycolytic process and protects from fibrosis [13].

Tuberculosis is one of the communicable diseases with a high mortality rate. It is caused by the bacteria mycobacterium tuberculosis. Usually, during infections, macrophages are activated to ingest and present these pathogens. Infection chronicity leads to the formation of granulomas that prevent the spread of the bacteria. However, a section of the population is highly susceptible to severe tuberculosis. This is because they have a defect in genes encoding receptors for IFN-γ on the surface of macrophage cells. As such, during infection, the macrophages are not activated to induce inflammation and present the pathogen for destruction but remain insensitive to IFN-y stimulation. The macrophages are thus unable to kill the bacteria. Moreover, the granulomas do not effectively form and systemic infection may occur.

In addition, macrophages contribute to diseases whose etiology does not primarily lie on defects associated with the activation, metabolism, and functioning of macrophages. These diseases include inflammatory diseases (inflammatory bowel disease, arthritis), infections (HIV), and metabolic diseases (such as atherosclerosis, obesity, and diabetes). The pathogenesis of such diseases usually utilizes the macrophage inflammatory process to develop its pathophysiology.

9. Conclusion

Macrophages are phagocytic immune cells. They function in both homeostasis and pathology. Macrophages are activated through two pathways depending on their polarization. Polarization refers to the ability of such cells to differentiate into functionally different phenotypes. Classically activated macrophages are also known as M1 macrophage, are pro inflammatory in nature and thus mediates inflammatory reactions. M2 macrophages on the other hand results from alternative activation of macrophages. The M2 macrophages are anti-inflammatory in nature functioning in wound healing and tissue repair. Macrophages can be classified based on their mode of activation and based on the tissue they are located in. based on activation, macrophages can fall into either M1 or M2 macrophages as stated above.

Different tissues have different types of macrophages depending on their roles and phenotypic expression. Kupffer cells in the liver, alveolar macrophages in the lungs, osteoclasts in bone tissue, Langerhans cells in the skin, intraocular macrophages in the eye, microglia in the brain, splenic macrophage, intestinal macrophage, and subscapular sinusoidal macrophages in lymph nodes. These macrophages have specific tissue functions in addition to the generalized functions of a macrophage cell. The generalized functions of a macrophage cell include induction of inflammation, immune surveillance, antigen processing, presentation wound healing, iron metabolism, and muscle regeneration. All these functions either fall into anti-inflammatory or proinflammatory. The mechanism of these functions is integrated with the metabolic profile of a macrophage and the type of polarization it acquires. M1 macrophages are activated mainly by IFN-γ and lipopolysaccharide while M2 macrophages are activated by interleukin 4. M1 macrophages primarily utilize aerobic glycolysis for energy production through inhibition of the Krebs cycle and cellular respiration. The resultant effect is the formation of metabolites such as itaconate with microbicidal effect and immediate provision of energy. In addition, M1 polarization results in the formation of reactive oxygen species due to inhibition of cellular respiration. These species get integrated into the phagosome to aid the phagocytic process. M2 macrophages on the other hand utilize fatty acid oxidation and cellular respiration to carry out their functions. Macrophages also have a role to play in the induction of pathological conditions. Primary defects in macrophage pathways result in the development of disorders such as increased susceptibility to tuberculosis, idiopathic pulmonary fibrosis, and macrophage activation syndrome. Secondarily, macrophages also aid in the pathogenesis of inflammatory diseases, infections, and metabolic syndromes.

10. Conflict of interest

The authors declare no conflict of interest.

Author details

Filex Otieno* and Cynthia Kyalo
Kabarak University, Nakuru, Kenya

*Address all correspondence to: filexotieno27@gmail.com

References

[1] Epelman S, Lavine K, Randolph G. Origin and functions of tissue macrophages. Immunity. 2014;**41**(1):21-35. DOI: 10.1016/j.immuni.2014.06.013

[2] Biswas SK, Mantovani A. Macrophages: Biology and Role in the Pathology of Diseases. Berlin: Springer; 2014

[3] Burke B, Lewis CE. The Macrophage. 2nd ed. Oxford: Oxford University Press; 2002

[4] Geissmann F, Manz MG, Jung S, Sieweke MH, Merad M, Ley K. Development of monocytes, macrophages, and dendritic cells. Science. 2010;**327**(5966):656-661. DOI: 10.1126/science.1178331

[5] Orecchioni M, Ghosheh Y, Pramod AB, Ley K. Macrophage polarization: Different gene signatures in M1(LPS+) vs. classically and m2(LPs–) vs. alternatively activated macrophages. Frontiers in Immunology. 2019;**10**(1084):1-14. DOI: 10.3389/fimmu.2019.01084

[6] Muraille E, Leo O, Moser M. Th1/Th2 paradigm extended: Macrophage polarization as an unappreciated pathogen-driven escape mechanism? Frontiers in Immunology. 2014;**5**(603):1-12. DOI: 10.3389/fimmu.2014.00603

[7] Biswas S, Mantovani A. Orchestration of metabolism by macrophages. Cell Metabolism. 2012;**15**(4):432-437. DOI: 10.1016/j.cmet.2011.11.013

[8] Classen A, Lloberas J, Celada A. Macrophage activation: Classical vs. Alternative. Macrophages and Dendritic Cells. 2009;**531**:29-43. DOI: 10.1007/978-1-59745-396-7_3

[9] Müller E, Speth M, Christopoulos PF, Lunde A, Avdagic A, Øynebråten I, et al. Both type I and type II interferons can activate antitumor M1 macrophages when combined with TLR stimulation. Frontiers in Immunology. 2018;**9**(2520):1-16. DOI: 10.3389/fimmu.2018.02520

[10] Weisz A, Oguchi S, Cicatiello L, Esumi H. Dual mechanism for the control of inducible-type NO synthase gene expression in macrophages during activation by interferon-gamma and bacterial lipopolysaccharide. Transcriptional and post-transcriptional regulation. Journal of Biological Chemistry. 1994;**269**(11):8324-8333. DOI: 10.1016/s0021-9258(17)37197-1

[11] Italiani P, Boraschi D. From monocytes to M1/M2 macrophages: Phenotypical vs. functional differentiation. Frontiers in Immunology. 2014;**5**(514):1-22. DOI: 10.3389/fimmu.2014.00514

[12] Gordon S, Martinez-Pomares L. Physiological roles of macrophages. PflügersArchiv—European Journal of Physiology. 2017;**469**(3-4):365-374. DOI: 10.1007/s00424-017-1945-7

[13] Wculek SK, Dunphy G, Heras-Murillo I, Mastrangelo A, Sancho D. Metabolism of tissue macrophages in homeostasis and pathology. Cellular & Molecular Immunology. 2021;**19**:384-408. DOI: 10.1038/s41423-021-00791-9

[14] Baranska A, Shawket A, Jouve M, Baratin M, Malosse C, Voluzan O, et al. Unveiling skin macrophage dynamics explains both tattoo persistence and strenuous removal. Journal of Experimental Medicine. 2018;**215**(4):1115-1133. DOI: 10.1084/jem.20171608

[15] Viola A, Munari F, Sánchez-Rodríguez R, Scolaro T, Castegna A. The metabolic signature of macrophage responses. Frontiers in Immunology. 2019;**10**(1462):1-16. DOI: 10.3389/fimmu.2019.01462

[16] Recalcati S, Gammella E, Cairo G. Ironing out macrophage immunometabolism. Pharmaceuticals. 2019;**12**(2):94. DOI: 10.3390/ph12020094

[17] Van den Bossche J, O'Neill LA, Menon D. Macrophage immunometabolism: Where are we (going)? Trends in Immunology. 2017;**38**(6):395-406. DOI: 10.1016/j.it.2017.03.001

[18] Ryan DG, O'Neill LA. Krebs cycle reborn in macrophage immunometabolism. Annual Review of Immunology. 2020;**38**(1):289-313. DOI: 10.1146/annurev-immunol-081619-104850

[19] Yang S, Van den Bossche J, Ramalho T. Macrophage metabolism at the crossroad of metabolic diseases and cancer. Immunometabolism. 2020;**2**(3):1-26. DOI: 10.20900/immunometab20200022

[20] Su X, Yu Y, Zhong Y, Giannopoulou EG, Hu X, Liu H, et al. Interferon-γ regulates cellular metabolism and mRNA translation to potentiate macrophage activation. Nature Immunology. 2015;**16**(8):838-849. DOI: 10.1038/ni.3205

[21] Meager A. The Interferons: Characterization and Application. Hoboken, New Jersey, U.S.: Wiley-Blackwell; 2006

[22] Wang F, Zhang S, Jeon R, Vuckovic I, Jiang X, Lerman A, et al. Interferon-gamma induces reversible metabolic reprogramming of M1 macrophages to sustain cell viability and pro-inflammatory activity. EBioMedicine. 2018;**30**:303-316. DOI: 10.1016/j.ebiom.2018.02.009

[23] Bories GF, Leitinger N. Macrophage metabolism in atherosclerosis. FEBS Letters. 2017;**591**(19):3042-3060. DOI: 10.1002/1873-3468.12786

[24] Al-Samkari H, Berliner N. HemophagocyticLymphohistiocytosis. Annual Review of Pathology: Mechanisms of Disease. 2018;**13**(1):27-49. DOI: 10.1146/annurev-pathol-020117-043625

[25] Grom AA. Genetics of macrophage activation syndrome in systemic juvenile idiopathic arthritis. Cytokine Storm Syndrome. 2019:131-138. DOI: 10.1007/978-3-030-22094-5_8

[26] Crayne CB, Albeituni S, Nichols KE, Cron RQ. The immunology of macrophage activation syndrome. Frontiers in Immunology. 2019;**10**(119):1-11. DOI: 10.3389/fimmu.2019.00119

[27] Schulert GS, Grom AA. Pathogenesis of macrophage activation syndrome and potential for cytokine—Directed therapies. Annual Review of Medicine. 2015;**66**(1):145-159. DOI: 10.1146/annurev-med-061813-012806

Section 2

Interferon in Therapy

Chapter 2

Interferons Horizon Therapeutics

Ayesha Aiman, Seemi Farhat Basir and Asimul Islam

Abstract

Interferons (IFNs) are a family of multi-functional proteins, called cytokines, that are produced by immune cells such as leukocytes, natural killer (NK) cells, macrophages, fibroblasts, and epithelial cells. The minute amount of these α-helical glycoproteins, produced by mammalian cells, are firm components of the innate arm of the immune system providing rapid and broad protection against numerous types of invading pathogens. Interferons, from their discovery in the 19th century, have always held out a promise of important clinical utility first as an antiviral agent and more recently holding anti-inflammatory and regenerative effects for treating various neurological diseases such as multiple sclerosis, encephalopathies, Alzheimer's disease (AD), Parkinson's disease (PD), amyotrophic lateral sclerosis (ALS), etc. IFNs elicit anti-viral and anti-inflammatory properties by inducing transcription of multiple IFN stimulated genes (ISG), a response that is partly mediated by Interferon regulatory factors (IRFs). This chapter provides a brief introduction of the interferon system as well as an in-depth assessment of the interferon signature and the various assay procedures for synthesizing non-natural interferon analogs for structural analysis, which may be helpful in designing improved products and act as a diagnostic tool for neurodegenerative disorders.

Keywords: cytokines, cancer, inflammation, interferons, interferon regulatory factors (IRFs), neurodegenerative disorders

1. Introduction

The name of the interferons comes from their capability to intrude with the product of new contagion patches. When the vulnerable system is attacked, they get actuated due to viral infection or other unknown substances and the white blood cells in the body produces interferons, which are a group of proteins called cytokines. Interferons (IFNs) are a group of soluble α-helical glycoproteins [1] that are produced and released by the innate arm of immune cells such as leukocytes, natural killer (NK) cells, fibroblasts, and epithelial cells in response to virus infection (or any other stimuli). They bind to specific IFN receptors on cells to trigger multiple signaling pathways that result in the expression of IFN-stimulated genes (ISGs). The ISG products then render the cell resistant to subsequent virus infection. Interferons do not directly kill the virus or cancerous cells rather they boost the vulnerable system response and reduce the growth of cancer cells by regulating the exertion of several genes that control the stashing of multitudinous cellular proteins that affect their growth. In extreme cases, an indecorous prosecution of these pathways or their inordinate activation can result in cell death. IFNs can play beneficial roles in the nervous system because of their tremendous capacity for upregulating immune responses. The three major types of interferons, that is, IFN-α, β and γ act as implicit curatives for a number of diseases. IFN-α has been used for the treatment of hepatitis B and C, and in several types of cancers, including hairy cell leukemia, chronic myeloid leukemia, Kaposi's sarcoma, and Erdheim–Chester disease (or polyostotic sclerotic histiocytosis), a rare complaint of bone marrow inflammation that can also affect the cerebellum [2]. IFN-β, an immunosuppressive cytokine, is the first medicine shown to promote clinical improvement in multiple sclerosis by inhibiting IL-12 production and inducing IL-10 [3]. Genetic research reveals the role of IFN-β in regulating mitochondrial dynamics to prevent neurodegeneration. IFN-β rescues mitochondrial abnormalities and neuronal survivance in Parkinson's disease *in vivo* [4]. Finally, IFN-γ has been used in the treatment of chronic granulomatous disease, a rare hereditable complaint in which the phagocytic cells have disabled capacity to kill ingested microbes, resulting in recurring bacterial and fungal infections [5]. IFNs, presumably in confluence with other cytokines, hold a prominent role in various therapies against diseases due to their incredible wide-ranging and pronounced immunological properties.

2. Origin and classification of interferons

Among the major discoveries in science, the discovery of interferon was a fortuitous one. The 60-year history of exploration on IFN abounds with big and small breakthroughs and are been recorded in the literature. However, information on the succession in lines of thought that led from one discovery to the next is dispersed, and many of those linkages may only be recorded in the memory of 'veteran' interferon workers. New generations of interferon workers tend to rely on handbooks or laboratory manuals, whereas background about sophisticated pathways of discovery is usually omitted. Therefore, this historical section related to molecular structure, production, and action of IFN will be considered from the viewpoint of how our insights have grown within the environment of evolving tools and general knowledge in cellular and molecular biology.

The basic phenomenon of interference was first described in the year 1935 with the capability of one contagion to interfere with the replication of another (challenge) contagion [6]. Thus, the hunt was underway for the mediator of viral interference for about 20 years until Alick Isaacs and Jean Lindenmann coined the term interferon (IFN) to it, in 1957 [7]. When heat- or UV-inactivated influenza virus was injected into the 10-day-old fragmented chorioallantoic membrane of chick embryos, a substance was released that inhibited viral multiplication. Hemagglutination, or the virus's ability to interact with and agglutinate red blood cells, was used to quantify influenza viral production (or inhibition). The interfering chemical was given the name "interferon." The titration ended when a well (on a plate of tiny wells) was identified with partial agglutination; the reciprocal of the influenza dilution thus measured was used as the interferon titer (concentration). Interferon molecules produced by infected cells function via autocrine and paracrine signaling to transform host cells into antiviral cells [8]. They have profound immunomodulatory as well as antiviral properties. They were initially classified as leukocyte, fibroblast, or immune IFNs, based on their cellular origin. Type I IFN (leukocyte and fibroblast IFN) and type II IFN (immune IFN) are two types of IFNs that are now known to be made up of over 20 distinct proteins [9].

3. IFNs family background

Even though IFNs were initially classified as antiviral agents, Isaac and Lindenmann could not have anticipated the enormous impact their discovery would have, and the extent to which they would be pertinent far beyond the discipline of Virology. From the discovery in the 1960s that IFNs also played a role in the control of cell growth and animal tumors up to the recent findings that they are pivotal regulators of both innate and adaptive immune responses, the result is that vertebrate life would be permanently threatened without IFNs.

Multiple criteria, including sequence identity, genetic loci, cell of origin, receptor distribution, and downstream reactions, have been used to classify IFNs. Although IFNs are expressed at low levels in the body at rest, they are activated to varying degrees depending on the stimuli, as a result, they play a dynamic and pathogen-specific function in the immune response. IFNs modulate the immune system's ability by promoting transcription of interferon signaling genes (ISGs) after they are generated and released by immune cells.

IFNs were classified as type-I (pH stable) or type-II (pH unstable) based on their pH sensitivity. The designation of IFN-α/β and IFN-γ as type-I and type-II IFNs, respectively, was further verified by analysis of their unique amino acid sequences and crystal structures [10–13]. The type-I family has been expanded to 16 members which include 12 IFN-αs that are encoded by 13 genes (IFN-α1/13 encode the same protein) [14–18], IFN-β (the well-known IFNs and the first to be cloned, purified, and sequenced) [17, 19, 20], IFN-ϵ [21], IFN-κ [22], and IFN-ω [23]. Type-II family includes only one member, that is, interferon- γ, produced by NK cells and T-cells (in response to cytokines IL-12 and IL-18). Both types of IFN promote an "anti-viral state" by snooping with cell proliferation and viral replication mechanisms. Moreover, IFNs render infected cells to become more susceptible to apoptosis (procaspase activity) and recognition by CD8+ cytotoxic T-cells by upregulating the expression of class I-major histocompatibility complex (MHC-I) on infected cells [24]. In 2003, the genome analysis discovered a novel type-III family of IFNs (IFN-λ), which were shown to be comparable to the IL-10 family of cytokines [16, 25–27], particularly IL-22 [28] based on sequence and subsequent structural studies. In humans, there are four different subtypes of Type III IFN, namely, IFN-λ1 (IL-29), IFN-λ2 (IL-28A), IFN-λ3 (IL-28B), and IFN-λ4 [29]. These IFN present similar biological effects to type-I IFNs, playing an important role in host defense against viral infections.

After a brief introduction of some of the cardinal features of the three types of interferons, we will now discuss the type of receptors involved in signal transduction pathways and biological activities elicited by them and then focus on the regulation of these IFN responses using transcription regulatory factors.

4. IFN induction and signaling mechanism

IFNs are incredibly effective at limiting virus replication and transmission, but because they are not normally expressed, IFN synthesis must be triggered promptly

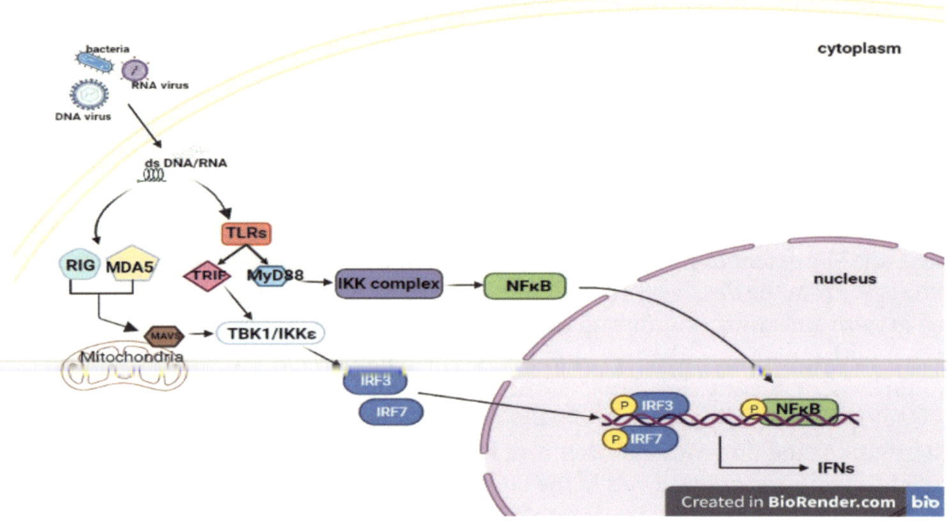

Figure 1.
The Mechanism of production of IFNs [30–32].

and strongly upon host contact with the virus. Because all viruses proliferate inside host cells, identifying bacterial or viral nucleic acids (e.g., RNA or DNA genome) upon microbial challenge, is an efficient technique for eliciting innate immune responses. These foreign substances are firstly identified by a specialized group of proteins known as Toll-like receptors (TLRs) which are, further, a type of pattern-recognition receptors (PRRs), that are expressed on sentinel cells. These receptors are either cytosolic or endosomal membrane proteins [30]. The binding of dsRNA/dsDNA to the helicase domain of RIG-I (Retinoic acid-inducible gene-I) and MDA5 (melanoma differentiation-associated gene-5), respectively, induces caspase activation following activation of tumor necrosis factor (TNF) receptor-associated factor (TRAF)-associated NF-κB activator, TANK binding kinase 1 (TBK1) and inhibitor of NF-κB kinase IKKε **Figure 1** [31–33]. IRF-3 and IRF-7 are expressed ubiquitously as inactive monomers in the cytosol but when cells are stimulated with poly (I:C) or virus infection, they get phosphorylated by the serine/threonine kinases, homodimerized, and are then translocated from cytosol to nucleus and binds to responsive elements for IFN-β gene transcription. After this, the secreted IFN binds to their specific cognate cell surface receptors, the heterodimeric IFNAR1/IFNAR2 complex for type I IFNs, dimers of the heterodimeric IFNGR1/IFNGR2 complex for type II IFN and the heterodimeric IFNLR1/IL10R2 complex for type III IFNs as represented in **Table 1** [32], present on the infected cell's surface, causing an autocrine signaling cascade that mobilizes other interferon response components and changes the gene expression patterns, resulting in an interferon response. IFNs can also bind to the interferon receptor produced by nearby non-virus infected cells, operating in a paracrine manner to enhance interferon response and aid these cells in combating viral infection [33].

The IFN signal transduction pathway has been appropriately described in multiple comprehensive reviews **Figure 2** [17, 29, 38, 40–42, 44–57]. The type I IFNs bind to their related heterodimeric cell surface receptors, IFNAR1 and IFNAR2, which signals through the activation of Janus activated kinases (JAKs), specifically TYK2 (Tyrosine kinase 2) and JAK1, respectively, causing tyrosine phosphorylation of the

Interferons	Discovery	Gene	Family members	Receptors	IFN- producing cells	Signalling	References
Type-I IFN	1957	Chr 9	IFN-α (13 subtypes), IFN-β, IFN-ε, IFN-κ, IFN-ω	IFNAR1/ IFNAR2 subunits	All nucleated cells	TYK2, JAK1, STAT1/2 forms ternary complex ISGF3, bind to ISRE/ GAS	[17, 34–38]
Type-II IFN	1965	Chr 12	IFN-γ	IFNGR1/ IFNGR2 subunits	B and T-lymphocytes, NK cells, APCs	JAK1/2, STAT1, bind to GAS/ISRE	[34, 35, 39–42]
Type-III IFN	2003	Chr 19	IFN-λ1, IFN-λ2, IFN-λ3, IFN-λ4	IFNLR1/ IL10R2 complex	All nucleated cells, dendritic cells, and epithelial cells	TYK2. JAK1, ISGF3, bind to ISRE	[29, 34, 35, 43–45]

Table 1.
Classification of interferons in humans.

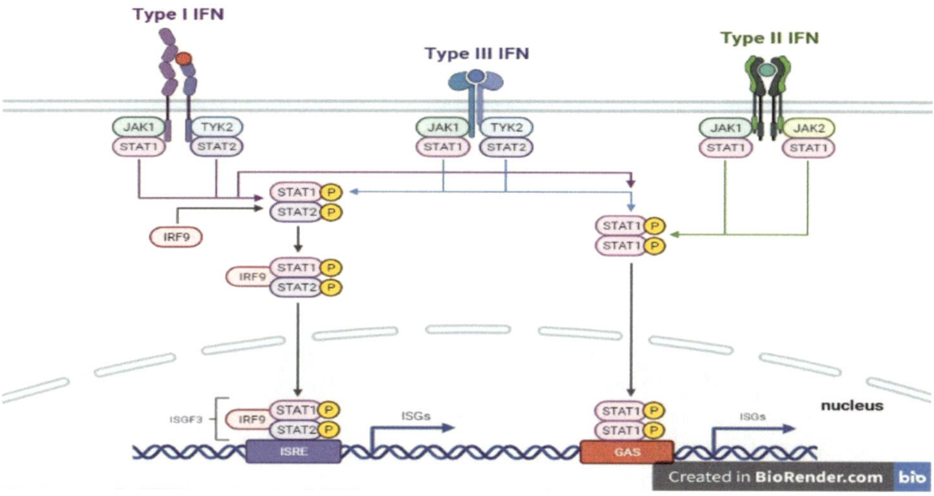

Figure 2.
Signal transduction mechanism by Type I, Type II and Type III IFN receptors and production of ISGs [17, 29, 33, 38–40, 42, 43, 45–55].

receptors' intracellular domains and recruitment of signal transducers and activator of transcription (STAT), STAT1 and STAT2 proteins, which in turn forms a trimeric complex, called ISGF3 (IFN stimulated gene factor 3) that consists of STAT1, STAT2, and IRF9 [17, 38]. The ISGF3 then translocate to the nucleus and binds to the IFN stimulated response element (ISRE) in the promoter region of IFN-stimulated genes (ISGs) and initiates transcription of antiviral genes. IRF2 acts as a transcriptional attenuator of ISGF3-mediated transcriptional activation within the nucleus, hence, the absence of IRF2 would result in increased Type I IFN signaling [46]. IFNAR activation also activates STAT1, STAT3, STAT4, STAT5, and STAT6 homodimers, as well as STAT1–STAT2, STAT1–STAT3, STAT1–STAT4, STAT1–STAT5, STAT2–STAT3 and STAT5–STAT6 heterodimers which bind and activates GAS (IFN-γ activated sequence) motifs, found in the promoter region of ISGs resulting in their gene expression [47–51]. Type I IFN signaling may also activate other signaling pathways that do not rely on the so-called JAK/STAT pathway. They are the non-canonical modifiers of Type I signaling called the mitogen-activated protein kinase (MAPK)/c-Jun amino-terminal kinase (JNK) pathways and the phosphoinositide 3-kinase (PI3K) pathway, which leads to diverse effects on the cell [52]. Furthermore, there is sufficient evidence that the function of distinct STATs may be modulated to account for individual responses. For example, a recent study found that STAT1 inhibits the IFN-α dependent induction of IFN-γ expression, whereas surprisingly, IFN-α or IFN-β mediated activation of STAT4 is essential for the IFN-γ synthesis during viral infection [53]. As a result, the functional diversity of type I IFN-regulated pathways allows for the transcriptional activation of a plethora of genes that facilitate the induction of physiologic responses.

Type II IFNs or IFN-γ is biologically active in its noncovalently coupled homodimer form. The extracellular domain of the two IFNGR1/CD119 subunit attaches to this homodimer, which then interacts with IFNGR2 to form a functional IFN- γ receptor complex. The receptor complex's IFNGR1 subunits are linked to JAK1, while the IFNGR2 subunits are linked to JAK2 [40]. When JAK1 and JAK2 are activated, the receptor is phosphorylated, and STAT1 is recruited and phosphorylated. The phosphorylation of STAT1 causes it to homodimerize and translocate to the nucleus. STAT1 homodimers attach to Gamma activated sequence (GAS) sites in the promoters of target genes once they reach the nucleus, regulating their transcription [41, 42]. IFN-γ signaling is dependent on weak type I IFN signaling, which is mediated by low type I IFN constitutive production [54]. Many of the IFN-gamma/STAT1 signaling-induced target genes are transcription factors that cause the expression of secondary response genes. IFN-gamma signaling can also activate the MAPK, PI3K/AKT/mTOR, and NF-kappa B signaling pathways, which control the expression of a variety of additional genes [55]. IFN-gamma signaling plays an important role in host defense by promoting macrophage activation, upregulating the expression of antigen processing and presentation molecules, driving the development and activation of Th1 cells, enhancing natural killer cell activity, regulating B cell functions, and inducing the production of chemokines that promote effector cell trafficking to sites of inflammation.

Type III interferons (IFN-λs) communicate with the body via a unique heterodimeric receptor complex, comprising of the IFN-λR1 subunit and interleukin-10R (IL-10R), shared by a number of cytokines in the IL-10 superfamily [29]. Despite the fact that IFN-λ and type I IFNs are structurally disparate and are engaged in different types of receptors, they both share the same JAK/STAT signal transduction pathway to trigger interferon-stimulated genes (ISGs), which have antiviral and immunoregulatory functions. So, these findings initially surmised that the Type I and Type III IFNs were functionally redundant. They were further distinguished by the kinetics of Type III interferons, which had a lower amplitude than Type I interferons while having long-lasting ISG expression [44, 45].

However, dysregulation of the IFN production and function would lead to immunological pathogenesis, such as inflammatory diseases, autoimmune and neurodegenerative disorders, via inappropriately stimulating inflammatory responses or dampening microbial controls. Thus, IFN response must be tightly regulated in order to develop protective immunity against microbial infections, curing autoimmune disorders and neurodegenerative diseases while avoiding detrimental toxicity induced by improper or prolonged gene expression.

5. IRF-mediated regulation of IFNs

Interferon regulatory factors (IRFs) are a group of transcription factors that are involved in a range of aspects of the innate and adaptive immune responses, including immune cell proliferation and differentiation, as well as modulating pathogenic responses [58]. Were first discovered and identified in the promoter region of the human interferon-β gene (IFN β1) during 1988, when a mouse cDNA clone encodes a protein that has specificity towards the IFN-β gene containing virus-inducible enhancer element, was identified [59]. During that period, there was no other homology present in accordance with this gene or other proteins. So, it was recognized and named as the IFN-regulatory factor 1 (IRF1). Further, cDNA clone that was identified later, subsequent cross-hybridization with IRF1 cDNA was named as IRF2. This signified the formal acknowledgment and birth of the massive IRF family [60]. IRFs specifically recognize the ISRE (Interferon-Stimulated Response Element), a conserved DNA consensus sequence, and become functionally active in the form of homodimers or heterodimers. The IRF family of transcription factors comprises of several members, namely, IRF1, IRF2, IRF3, IRF4 (also known as PIP, LSIRF or ICSAT), IRF5, IRF6, IRF7, IRF8 (also known as ICSBP), and IRF9 (also known as ISGF3γ/p48) were identified in Mus musculus and Homosapiens [61, 62]; IRF10 is observed in birds [63] and fishes [64], IRF11 is found in lower vertebrates, such as teleost fishes and zebrafish [65]. IRF1 and IRF2 have been extensively studied at the molecular level due to their unique properties of regulating gene expression despite having structural similarities. Although the former functions as a transcriptional activator and the latter repress IRF1 function by competing for the same cis-elements within type-I IFN (IFN-α/β) and IFN-inducible genes, they possess a high degree of structural similarity [66]. IRF-3, IRF-5, and IRF-7 are the three members of the IRF family which are induced by Type I IFNs, downstream of PRRs that detect viral DNA/RNA, resulting in a feedforward loop that maximally drives IFN expression [67]. Other members such as IRF-4, IRF-5, and IRF-8 are some of the key regulators of myeloid cell proliferation and phenotypic differentiation, which aids in modulating the inflammatory responses [68]. The fourth member of the family, IRF-9, controls the expression of a wide range of IFN stimulating genes. Researchers have discovered that, like the IFN-β gene, the IFN-λ1 gene is controlled by both IRF3 and IRF7, but IRF7 is the primary regulator of the IFN-λ2/3 genes [69, 70]. Understanding how their levels and activity are controlled is crucial since changes in either can lead to dysregulated immune responses and the development of autoimmune and neurodegenerative diseases.

6. Interferon system dysfunction and related disorders

Interferons are used to treat a number of diseases such as those that are caused by viruses (such as hepatitis B and C virus) or due to inflammation (like multiple sclerosis and systemic sclerosis) as depicted in **Table 1** [34, 71, 72]. They also act as antineoplastic agents to treat malignancies (such as breast carcinoma, nodular lymphoma, chronic myelogenous leukemia (CML), Kaposi's sarcoma, and renal adenocarcinoma) [73]. The expression levels of IFNs, as well as their actions, are superbly controlled in order to protect host cells from potential toxicity resulting from excessive responses. However, persistent and dysregulated IFN expression causes many diseases such as Type I interferonopathy, a type of inherited CNS disease. They have also been associated with the development or worsening of autoimmune diseases such as psoriasis, systemic lupus erythematosus (SLE), and, in rare cases, rheumatoid arthritis (RA) [74]. This was observed in their mRNA expression patterns that contain the interferon signature. Moreover, a lot of murine Alzheimer's disease (AD) models, as well as wild-type mouse brains challenged with generic nucleic acid-containing amyloid fibrils, showed an increased IFN-stimulated gene (ISG) signature [75]. These findings all point to IFNs that have a negative impact on the brain. Interferon overactivation is also associated with low levels of apoptotic particle clearance, resulting in an accumulation of apoptotic products (such as DNA-CpG motifs and U-RNAs). Similar abnormalities are seen in patients with primary Sjogren's syndrome, systemic scleroderma, and polymyositis, as well as a few cases of rheumatoid arthritis. Immunomodulation treatments aiming at lowering interferon overactivity are being tried in people with such diseases [76].

Shreds of evidence have shown that the development of autoimmune illnesses in certain people who were given IFN-α suggests that this cytokine plays a key role in breaking tolerance and triggering autoimmune responses in such patients [77]. Similarly, IFN-γ may also contribute to autoimmune disorders in addition to its host defense actions. Although IFN-γ production has been reported to be disease-limiting in experimental allergic encephalomyelitis (EAE), it may have a role in autoimmune nephritis [78]. Moreover, increased vulnerability to infection with certain viruses and intracellular bacteria appears to be linked to the loss of functioning IFNGR1 that is involved in Type II IFN signaling [79].

Recent research suggests that even mild-to-moderate acute COVID-19 infection results in a continuing, prolonged inflammatory response, which is not seen with common coronavirus infection [80]. After surviving acute coronavirus disease 2019 (COVID-19) infection, some individuals develop post-acute COVID syndrome (long COVID (LC)) that lasts longer than 12 weeks. The mechanisms behind this activation are still being investigated, but they might include antigen persistence, autoimmunity triggered by antigenic cross-reactivity, or a reflection of damage repair. These findings show that individuals with COVID-19 have an aberrant immunological profile at long intervals after infection, indicating the presence of an LC syndrome [80]. In this aspect, learning more about the immunological components of diverse pathologies has yielded common themes. Because of these unifying concepts, immune-based therapeutics for viral respiratory diseases, autoimmune and neurodegenerative disorders must be identified.

7. Closing remarks and outlook

There is a worldwide interest in repurposing existing drugs and understanding mechanisms against viral, autoimmune, and neurodegenerative diseases. Structural determination, interaction with different co-solutes, and binding studies can facilitate the process of vaccine development, help in understanding the mechanism of anti-inflammation, and design a potent inhibitor for drug discovery. The interaction studies with different proteins will stabilize and/or destabilize, allowing deeper insight into various interactions (attractive and repulsive forces) to maintain a high functional protein population as this can probably be helpful for pre-clinical toxicological studies. Furthermore, beyond the therapeutic benefit to the individual patient, IFN therapy may aid public health measures aimed at delaying the spread of pandemic diseases and also minimizing the deterioration of symptoms in cases of autoimmune diseases and neurodegenerative disorders by reducing the time it takes for their symptoms to deteriorate. However, the most difficult element of creating therapy options for immune modulation against such illnesses is disentangling beneficial from harmful signals. So, for that purpose, targeted immune regulation can temper maladaptive factors enabling beneficial immune response against disorders which might help reduce its severity in the future.

Author details

Ayesha Aiman[1], Seemi Farhat Basir[1]* and Asimul Islam[2]*

1 Department of Biosciences, Jamia Millia Islamia, New Delhi, India

2 Centre for Interdisciplinary Research in Basic Sciences, Jamia Millia Islamia, New Delhi, India

*Address all correspondence to: sbasir@jmi.ac.in and aislam@jmi.ac.in

References

[1] Klaus W, Gsell B, Labhardt AM, Wipf B, Senn H. The three-dimensional high resolution structure of human interferon α-2a determined by heteronuclear NMR spectroscopy in solution. Journal of Molecular Biology. 1997;**274**(4):661-675. DOI: 10.1006/jmbi.1997.1396

[2] Selmi C, Lleo A, Zuin M, Podda M, Rossaro L, Gershwin ME. Interferon α and its contribution to autoimmunity. Current Opinion in Investigational Drugs. 2006;**7**(5):451-456. Accessed: May 30, 2022. [Online]. Available from: https://europepmc.org/article/med/16729722

[3] Graber JJ, Ford D, Zhan M, Francis G, Panitch H, Dhib-Jalbut S. Cytokine changes during interferon-beta therapy in multiple sclerosis: Correlations with interferon dose and MRI response. Journal of Neuroimmunology. 2007;**185**(1-2):168-174. DOI: 10.1016/j.jneuroim.2007.01.011

[4] Tresse E, Riera-Ponsati L, Jaberi E, Sew WQG, Ruscher K, Issazadeh-Navikas S. IFN-β rescues neurodegeneration by regulating mitochondrial fission via STAT5, PGAM5, and Drp1. The EMBO Journal. 2021;**40**(11). DOI: 10.15252/embj.2020106868

[5] Åhlin A, Lärfars G, Elinder G, Palmblad J, Gyllenhammar H. Gamma interferon treatment of patients with chronic granulomatous disease is associated with augmented production of nitric oxide by polymorphonuclear neutrophils. Clinical and Diagnostic Laboratory Immunology. 1999;**6**(3):420-424. DOI: 10.1128/cdli.6.3.420-424.1999

[6] Hoskins M. A Protective Action of Neurotropic Against Viscerotropic Yellow Fever Virus in Macacus Rhesus. The American Journal of Tropical Medicine and Hygiene. 1935;**s1-15**(6):675-680. DOI: 10.4269/ajtmh.1935.s1-15.675

[7] Virus interference. I. The interferon. Proceedings of the Royal Society B: Biological Sciences. 1957;**147**(927):258-267. DOI: 10.1098/rspb.1957.0048

[8] Michael Lavigne G, Russell H, Sherry B, Ke R. Autocrine and paracrine interferon signalling as 'ring vaccination' and 'contact tracing' strategies to suppress virus infection in a host. Proceedings of the Royal Society B: Biological Sciences. 2021;**288**(1945). DOI: 10.1098/rspb.2020.3002

[9] Interferons. Multiple Sclerosis Journal. 1999:259-292. DOI: 10.1201/9781439812242-31

[10] Radhakrishnan R et al. Zinc mediated dimer of human interferon-α(2b) revealed by X-ray crystallography. Structure. 1996;**4**(12):1453-1463. DOI: 10.1016/S0969-2126(96)00152-9

[11] Karpusas M, Nolte M, Benton CB, Meier W, Lipscomb WN, Goelz S. The crystal structure of human interferon β at 2.2-Å resolution. Proceedings of the National Academy of Sciences of the United States of America. 1997;**94**(22):11813-11818. DOI: 10.1073/pnas.94.22.11813

[12] Ealick SE et al. Three-dimensional structure of recombinant human interferon-γ. Science (80-.). 1991;**252**(5006):698-702. DOI: 10.1126/science.1902591

[13] Landar A et al. Design, characterization, and structure of a biologically active single-chain mutant

of human IFN-γ. Journal of Molecular Biology. 2000;**299**(1):169-179. DOI: 10.1006/jmbi.2000.3734

[14] Pestka S. The human interferon-alpha species and hybrid proteins. Seminars in Oncology. 1997;**24**(3 Suppl 9). Accessed: May 30, 2022. [Online]. Available from: https://europepmc.org/article/med/9208870

[15] Pestka S. The Interferons: 50 Years after their discovery, there is much more to learn. Journal of Biological Chemistry. 2007;**282**(28):20047-20051. DOI: 10.1074/jbc.R700004200

[16] Pestka S, Krause CD, Sarkar D, Walter MR, Shi Y, Fisher PB. Interleukin-10 and related cytokines and receptors. Annual Review of Immunology. 2004;**22**:929-979. DOI: 10.1146/annurev.immunol.22.012703.104622

[17] Pestka S, Krause CD, Walter MR. Interferons, interferon-like cytokines, and their receptors. Immunological Reviews. 2004;**202**:8-32. DOI: 10.1111/j.0105-2896.2004.00204.x

[18] Kuruganti S, Accavitti-Loper MA, Walter MR. Production and characterization of thirteen human type-I interferon-α subtypes. Protein Expression and Purification. 2014;**103**:75-83. DOI: 10.1016/j.pep.2014.08.010

[19] Taniguchi T, Fujii-Kuriyama Y, Muramatsu M. Molecular cloning of human interferon cDNA. Proceedings of the National Academy of Sciences of the United States of America. 1980;**77**(7 II):4003-4006. DOI: 10.1073/pnas.77.7.4003

[20] Stark GR, Kerr IM, Williams BRG, Silverman RH, Schreiber RD. How cells respond to interferons. Annual Review of Biochemistry. 1998;**67**:227-264. DOI: 10.1146/annurev.biochem.67.1.227

[21] Hardy MP, Owczarek CM, Jermiin LS, Ejdebäck M, Hertzog PJ. Characterization of the type I interferon locus and identification of novel genes. Genomics. 2004;**84**(2):331-345. DOI: 10.1016/j.ygeno.2004.03.003

[22] LaFleur DW et al. Interferon-κ, a Novel Type I Interferon Expressed in Human Keratinocytes. The Journal of Biological Chemistry. 2001;**276**(43):39765-39771. DOI. 10.1074/jbc.M102502200

[23] Hauptmann R, P. S.-N. acids research, and undefined 1985. A novel class of human type I interferons. academic.oup.com. 1985;**13**. Accessed: May 30, 2022. [Online]. Available from: https://academic.oup.com/nar/article-abstract/13/13/4739/1119648

[24] Zhou F. Molecular mechanisms of IFN-γ to up-regulate MHC class i antigen processing and presentation. International Reviews of Immunology. 2009;**28**(3-4):239-260. DOI: 10.1080/08830180902978120

[25] Walter MR. Structure of interleukin-10/interleukin-10R1 complex: A paradigm for class 2 cytokine activation. Immunologic Research. 2002;**26**(1-3):303-308. DOI: 10.1385/IR:26:1-3:303

[26] Walter MR. Structural analysis of IL-10 and type I interferon family members and their complexes with receptor. Advances in Protein Chemistry. 2004;**68**:171-223. DOI: 10.1016/S0065-3233(04)68006-5

[27] Ouyang W, Rutz S, Crellin NK, Valdez PA, Hymowitz SG. Regulation and functions of the IL-10 family of cytokines in inflammation and disease. Annual Review of Immunology. 2011;**29**:71-109. DOI: 10.1146/ANNUREV-IMMUNOL-031210-101312

[28] Nagem RAP, Colau D, Dumoutier L, Renauld JC, Ogata C, Polikarpov I. Crystal structure of recombinant human interleukin-22. Structure. 2002;**10**(8):1051-1062. DOI: 10.1016/S0969-2126(02)00797-9

[29] Kotenko SV et al. IFN-λs mediate antiviral protection through a distinct class II cytokine receptor complex. Nature Immunology. 2003;**4**(1):69-77. DOI: 10.1038/ni875

[30] Mielcarska MB, Bossowska-Nowicka M, Toka FN. Cell surface expression of endosomal toll-like receptors—A necessity or a superfluous duplication? Frontiers in Immunology. 2021;**11**. DOI: 10.3389/FIMMU.2020.620972/FULL

[31] Negishi H, Taniguchi T, Yanai H. The interferon (IFN) class of cytokines and the IFN regulatory factor (IRF) transcription factor family. Cold Spring Harbor Perspectives in Biology. 2018;**10**(11). DOI: 10.1101/cshperspect.a028423

[32] Walter MR. The role of structure in the biology of interferon signaling. Frontiers in Immunology. 2020;**11**. DOI: 10.3389/FIMMU.2020.606489/FULL

[33] Stetson DB, Medzhitov R. Type I interferons in host defense. Immunity. 2006;**25**(3):373-381. DOI: 10.1016/j.immuni.2006.08.007

[34] Mahajan BB, Kaur S. Interferons. Indian Journal of Dermatology, Venereology and Leprology. 2015;**81**(1):51-55. DOI: 10.4103/0378-6323.148570

[35] Tan G, Song H, Xu F, Cheng G. When hepatitis B virus meets interferons. Frontiers in Microbiology. 2018;**9**(JUL). Frontiers Media S.A. DOI: 10.3389/fmicb.2018.01611

[36] Isaacs A, Lindenmann J. Virus interference. I. The interferon. Journal of Interferon Research. 1987;**7**(5):429-438. DOI: 10.1089/jir.1987.7.429

[37] Bekisz J, Schmeisser H, Hernandez J, Goldman ND, Zoon KC. Human interferons alpha, beta and omega. Growth Factors. 2004;**22**(4):243-251. DOI: 10.1080/08977190400000833

[38] Ivashkiv LB, Donlin LT. Regulation of type i interferon responses. Nature Reviews Immunology. 2014;**14**(1):36-49. DOI: 10.1038/nri3581

[39] Frederick Wheelock E. Interferon-like virus-inhibitor induced in human leukocytes by phytohemagglutinin. Science (80-.). 1965;**149**(3681):310-311. DOI: 10.1126/science.149.3681.310

[40] Farrar MA, Schreiber RD. The molecular cell biology of interferon-γ and its receptor. Annual Review of Immunology. 1993;**11**:571-611. DOI: 10.1146/ANNUREV.IY.11.040193.003035

[41] Butturini E, de Prati AC, Mariotto S. Redox regulation of STAT1 and STAT3 signaling. International Journal of Molecular Sciences. 2020;**21**(19):1-18. DOI: 10.3390/ijms21197034

[42] Alspach E, Lussier DM, Schreiber RD. Interferon γ and its important roles in promoting and inhibiting spontaneous and therapeutic cancer immunity. Cold Spring Harbor Perspectives in Biology. 2019;**11**(3). DOI: 10.1101/cshperspect.a028480

[43] Sheppard P, Kindsvogel W, Xu W, et al. IL-28, IL-29 and their class II cytokine receptor IL-28R. nature.com, 2003. 2003. DOI: 10.1038/ni873

[44] Goel RR, Kotenko SV, Kaplan MJ. Interferon lambda in inflammation and autoimmune rheumatic diseases. Nature Reviews Rheumatology. 2021;**17**(6):349-362. DOI: 10.1038/s41584-021-00606-1

[45] Wack A, Terczyńska-Dyla E, Hartmann R. Guarding the frontiers: The biology of type III interferons. Nature Immunology. 2015;**16**(8):802-809. DOI: 10.1038/ni.3212

[46] Iida S et al. CD8+ T cell-mediated skin disease in mice lacking IRF-2, the transcriptional attenuator of interferon-α/β signaling. Immunity. 2000;**13**(5):643-655. DOI: 10.1016/S1074-7613(00)00064-9

[47] Decker T, Müller M, Stockinger S. The Yin and Yang of type I interferon activity in bacterial infection. Nature Reviews Immunology. 2005;**5**(9):675-687. DOI: 10.1038/nri1684

[48] Platanias LC, Fish EN. Signaling pathways activated by interferons. Experimental Hematology. 1999;**27**(11):1583-1592. DOI: 10.1016/S0301-472X(99)00109-5

[49] Parmar S, Platanias LC. Interferons: Mechanisms of action and clinical applications. Current Opinion in Oncology. 2003;**15**(6):431-439. DOI: 10.1097/00001622-200311000-00005

[50] Darnell JE. STATs and gene regulation. Science (80-.). 1997;**277**(5332):1630-1635. DOI: 10.1126/science.277.5332.1630

[51] K. Kishimoto et al., "High dose IFN-β activates GAF to enhance expression of ISGF3 target genes in MLE12 epithelial cells," Frontiers in Immunology, vol. 12, Apr. 2021, doi: 10.3389/fimmu.2021.651254.

[52] Mazewski C, Perez RE, Fish EN, Platanias LC. Type I interferon (IFN)-regulated activation of canonical and non-canonical signaling pathways. Frontiers in Immunology. 2020;**11**. Frontiers Media S.A. DOI: 10.3389/fimmu.2020.606456

[53] Nguyen KB et al. Critical role for STAT4 activation by type 1 interferons in the interferon-γ response to viral infection. Science (80-.). 2002;**297**(5589):2063-2066. DOI: 10.1126/science.1071900

[54] Takaoka A et al. Cross talk between interferon-γ and -α/β signaling components in caveolar membrane domains. Science (80-.). 2000;**288**(5475):2357-2360. DOI: 10.1126/science.288.5475.2357

[55] Green DS, Young HA, Valencia JC. Current prospects of type II interferon γ signaling & autoimmunity. Journal of Biological Chemistry. 2017;**292**(34):13925-13933. DOI: 10.1074/jbc.R116.774745

[56] Haque SJ, Williams BRG. Signal transduction in the interferon system. Seminars in Oncology. 1998;**25**(1 SUPPL):14-22. Accessed: May 30, 2022. [Online]. Available from: https://europepmc.org/article/med/9482536

[57] Chen K, Liu J, Cao X. Regulation of type I interferon signaling in immunity and inflammation: A comprehensive review. Journal of Autoimmunity. 2017;**83**:1-11. DOI: 10.1016/j.jaut.2017.03.008

[58] Tamura T, Yanai H, Savitsky D, Taniguchi T. The IRF family transcription factors in immunity and oncogenesis. Annual Review of Immunology. 2008;**26**:535-584. DOI: 10.1146/annurev.immunol.26.021607.090400

[59] Miyamoto M et al. Regulated expression of a gene encoding a nuclear factor, IRF-1, that specifically binds to IFN-β gene regulatory elements. Cell. 1988;**54**(6):903-913. DOI: 10.1016/S0092-8674(88)91307-4

[60] Harada H et al. Structurally similar but functionally distinct factors, IRF-1

and IRF-2, bind to the same regulatory elements of IFN and IFN-inducible genes. Cell. 1989;**58**(4):729-739. DOI: 10.1016/0092-8674(89)90107-4

[61] Y. Mamane et al., "Interferon regulatory factors: the next generation.," Gene, vol. 237, pp. 1-14, 1999. Accessed: May 31, 2022. [Online]. Available from: https://www.sciencedirect.com/science/article/pii/S0378111999002620

[62] Taniguchi T, Ogasawara K, Takaoka A, Tanaka N. IRF family of transcription factors as regulators of host defense. Annual Review of Immunology. 2001;**19**:623-655. DOI: 10.1146/ANNUREV.IMMUNOL.19.1.623

[63] Nehyba J, Hrdličková R, Burnside J, Bose HR. A novel interferon regulatory factor (IRF), IRF-10, has a unique role in immune defense and is induced by the v-Rel oncoprotein. Molecular and Cellular Biology. 2002;**22**(11):3942-3957. DOI: 10.1128/MCB.22.11.3942-3957.2002

[64] Xu Q et al. Sequence and expression analysis of interferon regulatory factor 10 (IRF10) in three diverse teleost fish reveals its role in antiviral defense. PLoS One. 2016;**11**(1). DOI: 10.1371/JOURNAL.PONE.0147181

[65] Huang B, Qi ZT, Xu Z, Nie P. Global characterization of interferon regulatory factor (IRF) genes in vertebrates: Glimpse of the diversification in evolution. BMC Immunology. 2010;**11**. DOI: 10.1186/1471-2172-11-22

[66] H. Harada, E. Takahashi, S. Itoh, K. Harada, T. A. Hori, and T. Taniguchi, "Structure and regulation of the human interferon regulatory factor 1 (IRF-1) and IRF-2 genes: implications for a gene network in the interferon system," Molecular and Cellular Biology, vol. 14, no. 2, pp. 1500-1509, Feb. 1994, doi: 10.1128/MCB.14.2.1500-1509.1994.

[67] Barnes B, Lubyova B, Pitha PM. On the role of IRF in host defense. Journal of Interferon and Cytokine Research. 2002;**22**(1):59-71. DOI: 10.1089/107999002753452665

[68] Jefferies CA. Regulating IRFs in IFN driven disease. Frontiers in Immunology. 2019;**10**(MAR). Frontiers Media S.A. DOI: 10.3389/fimmu.2019.00325

[69] Paul A, Tang TH, Ng SK. Interferon regulatory factor 9 structure and regulation. Frontiers in Immunology. 2018;**9**(AUG). Frontiers Media S.A. DOI: 10.3389/fimmu.2018.01831

[70] Österlund PI, Pietilä TE, Veckman V, Kotenko SV, Julkunen I. IFN regulatory factor family members differentially regulate the expression of type III IFN (IFN-λ) genes. Journal of Immunology. 2007;**179**(6):3434-3442. DOI: 10.4049/jimmunol.179.6.3434

[71] Maher S, Romero-Weaver A, Scarzello A, Gamero A. Interferon: Cellular executioner or white knight? Current Medicinal Chemistry. 2007;**14**(12):1279-1289. DOI: 10.2174/092986707780597907

[72] Wittling MC, Cahalan SR, Levenson EA, Rabin RL. Shared and unique features of human interferon-beta and interferon-alpha subtypes. Frontiers in Immunology. 2021;**11**. Frontiers Media S.A. DOI: 10.3389/fimmu.2020.605673

[73] Abdolvahab MH, Darvishi B, Zarei M, Majidzadeh-A K, Farahmand L. Interferons: Role in cancer therapy. Immunotherapy. 2020;**12**(11). Future Medicine Ltd.:833-855. DOI: 10.2217/imt-2019-0217

[74] Yasmine M et al. THU0606 rheumatoid arthritis induced by alpha-interferon therapy: A rare case

presentation. Annals of the Rheumatic Diseases. 2020;**79**(Suppl 1): 544.1-544.54545. DOI: 10.1136/annrheumdis-2020-eular.2286

[75] Roy ER et al. Type I interferon response drives neuroinflammation and synapse loss in Alzheimer disease. The Journal of Clinical Investigation. 2020;**130**(4):1912-1930. DOI: 10.1172/JCI133737

[76] Meyer O. Interferons and autoimmune disorders. Joint, Bone, Spine. 2009;**76**(5):464-473. DOI: 10.1016/j.jbspin.2009.03.012

[77] Rönnblom L. The type i interferon system in the etiopathogenesis of autoimmune diseases. Upsala Journal of Medical Sciences. 2011;**116**(4):227-237. DOI: 10.3109/03009734.2011.624649

[78] Espejo C et al. Treatment with anti-interferon-γ monoclonal antibodies modifies experimental autoimmune encephalomyelitis in interferon-γ receptor knockout mice. Experimental Neurology. 2001;**172**(2):460-468. DOI: 10.1006/exnr.2001.7815

[79] Schroder K, Hertzog PJ, Ravasi T, Hume DA. Interferon-γ: an overview of signals, mechanisms and functions. Journal of Leukocyte Biology. 2004;**75**(2):163-189. DOI: 10.1189/jlb.0603252

[80] Phetsouphanh C et al. Immunological dysfunction persists for 8 months following initial mild-to-moderate SARS-CoV-2 infection. Nature Immunology. 2022;**23**(2):210-216. DOI: 10.1038/s41590-021-01113-x

Chapter 3

Perspective Chapter: Impact of Interferon Alpha/Beta in the Management of Chronic Myeloproliferative Disorders

Anna Prochwicz and Dorota Krochmalczyk

Abstract

It has been noted that interferon can exert an antiproliferative effect by stimulating cells of the immune system. Interferon has been shown to be effective in the treatment of chronic myeloproliferative neoplasms. Over the years, interferon alpha-2a and interferon alpha-2b have been introduced into the treatment of chronic myeloproliferation, followed by their pegylated forms. Studies have been showing the effectiveness of interferon alpha in reducing the number of platelets in essential thrombocythemia, reducing the need for phlebotomies in patients with polycythemia vera and also in reducing the number of leukocytes. Additionally, it turned out to be effective in reducing the size of the spleen. Interferon has also been shown to be effective in inducing long-term molecular remissions. The introduction of new forms of interferon such as ropeginterferon and the combination of interferon alpha with newly introduced drugs from other groups causes that interferon remains an important drug in the field of chronic myeloproliferative disorders. The chapter presents the results of clinical trials and the experiences of various centers in its usage for mieloproliferative neoplasms.

Keywords: chronic myeloproliferative disorders, polycythemia vera, essential thrombocythemia, myelofibrosis, ropeginterferon, pegylated interferon alpha

1. Introduction

Chronic myeloproliferative disorders are a group of clonal diseases of the stem cell. It is a group of several diseases with some common features. They derive from a multipotential hematopoietic stem cell. A clone of neoplastic cells in all these neoplams is characterized by a lower proliferative activity than that of acute myeloproliferative diseases. In each of these diseases, leukocytosis, thrombocythemia, and polyglobulia may appear at some stage, depending on the diagnosis [1, 2].

The research on interferon has been going on since the 1950s [3]. Then, the attention was paid to its influence on the immune system. It has been noted that it can exert an antiproliferative effect by stimulating cells of the immune system [4]. In

1987, a publication by Ludwig et al. was published, which reported the effectiveness of interferon alpha in the treatment of chronic myeloproliferative disorders [5].

More and more new studies have been showing the effectiveness of interferon alpha in reducing the number of platelets, reducing the need for phlebotomies in patients with polycythemia vera and also in reducing the number of leukocytes. Moreover, interferon reduced the symptoms of myeloproliferative disorders such as redness and itching of the skin. Additionally, it turned out to be effective in reducing the size of the spleen.

Further studies on the assessment of remission using molecular-level response assessments indicate that the interferon action in chronic myeloproliferation diseases targets cells from the mutant clone with no effect on normal bone marrow cells [6].

Over the years, interferon alpha-2a and interferon alpha-2b have been introduced into the treatment of chronic myeloproliferation, followed by their pegylated forms. The introduction of pegylated forms allowed for a reduction in the number of side effects and less frequent administration of the drug to patients. In recent years, monopegylated interferon alpha-2b has been used to further increase the interval between drug administrations while maintaining its antiproliferative efficacy.

The exact mechanism of action of interferon alpha in the treatment of chronic myeloproliferative disease is still not fully understood, but it has an impact on JAK2 (Janus Kinase) signal transducers and activates the STAT signal pathway (Janus Kinase/SignalTransducer and Activator of Transcription).

Interferon alpha binds to IFNAR1 and IFNAR2c, which are type I interferon receptors. Interferon alpha has an impact on JAK2(Janus Kinase) signal transducers and activates the STAT signal pathway. The disturbances in this signaling pathway are observed in chronic myeloproliferative disorders [7].

Interferon inhibits the JAK-STAT signaling pathway by directly inhibiting the action of thrombopoietin in this pathway [8].

So far, three driver mutations have been described in the course of chronic myeloproliferative diseases that affect the functioning of the JAK-STAT pathway.

JAK2 kinase and JAK1, JAK3, and TYK2 kinases belong to the family of non-receptor tyrosine kinases. They are involved in the intracellular signal transduction of the JAK-STAT pathway. It is a system of intracellular proteins used by growth factors and cytokines to express genes that regulate cell activation, proliferation, and differentiation. The mechanism of JAK activation is based on the autophosphorylation of tyrosine residues that occurs after ligand binds to the receptor. JAK2 kinase transmits signals from the hematopoietic cytokine receptors of the myeloid lineage (erythropoietin, granulocyte-colony stimulating factor thrombopoietin, and lymphoid lineage [9].

A somatic G/T point mutation in exon 14 of the JAK2 kinase gene converts valine to phenylalanine at position 617 (V617F) in the JAK2 pseudokinase domain, which allows constitutive, ligand-independent activation of the receptor to trigger a proliferative signal [10].

Mutation of the MPL gene, which encodes the receptor for thrombopoietin, increases the sensitivity of magekaryocytes to the action of thrombopoietin, which stimulates their proliferation [11].

Malfunction of calreticulin as a result of mutation of the CARL gene leads to the activation of the MPL-JAK/STAT signaling pathway, which is independent of the ligand, as calreticulin is responsible, for the proper formation of the MPL receptor. Consequently, there is a clonal proliferation of hematopoietic stem cells [12].

Below, we provide an overview of some clinical studies on the efficacy of interferon in chronic myeloproliferative disorders.

2. Chronic myeloproliferative disorders

2.1 Polycythemia vera

Polycythemia vera (PV) is characterized by an increase in the number of erythrocytes in the peripheral blood.

Polycythemia vera is caused by a clonal mutation in the multipotential hematopoietic stem cell of the bone marrow. The mutation leads to an uncontrolled proliferation of the mutated cell clone, independent of erythropoietin and other regulatory factors. As the mutation takes place at an early stage of hematopoiesis, an increase of the number of erythrocytes as well as of leukocytes and platelets is observed in the peripheral blood. The cause of proliferation in PV independent from external factors is a mutation in the Janus 2 (JAK2) tyrosine kinase gene. The V617F point mutation in the JAK2 gene is responsible for about 96% mutation, and in the remaining cases the mutation arises in exon 12. Both mutations lead to constitutive activation of the JAK-STAT signaling pathway [13].

As a result of the uncontrolled proliferation, blood viscosity increases, which generates symptoms such as headaches and dizziness, visual disturbances, or erythromelalgia. As the number of all hematopoietic cells, including the granulocytes ones, increases, the difficult to control symptoms of their hyperdegranulation may appear, among which gastric ulcer or skin itching is often observed. During the disease progression, the spleen and liver become enlarged.

The most common complication of the disease is episodes of thrombosis, especially arterial one. During the course of the disease, it can also evolve into myelofibrosis or acute myeloid leukemia.

The treatment of PV is aimed at preventing thromboembolic complications, relieving the general symptoms, the appearance of hepatosplenomegaly as well as preventing its progression.

Each patient should receive an antiplatelet drug chronically, and usually acetylsalicylic acid is the choice. Most often, the treatment is started with phlebotomy in order to rapidly lower the hematocrit level. If cytoreductive therapy is necessary, the drugs of first choice are hydroxycarbamide and interferon [2].

However, the research on the mechanism of the action of interferons is still ongoing. In vitro studies with CD34+ cells from peripheral blood of patients diagnosed with polycythemia vera showed that interferon inhibits clonal changed cells selectively. It was found that interferon alpha-2b and pegylated interferon alpha-2a reduce the percentage of cells with JAK2 V617F mutation by about 40%. Pegylated interferon alpha-2a works by activating mitogen-activated protein kinase P38. It affects CD34+ cells of patients with polycythemia vera by increasing the rate of their apoptosis [6].

A case of a patient with PV with a confirmed chromosomal translocation t(6;8) treated with interferon alpha-2b, which resulted in a reduction of the clone with translocation by 50% from the baseline value, was also described [14].

In 2019, the results of a phase II multicenter study were published, which aimed at assessing the effectiveness of recombinant pegylated interferon alpha-2a in cases of refractory to previously hydroxycarbamide therapy. The study included 65 patients with essential thrombocythemia (ET) and 50 patients with polycythemia vera. All patients had previously been treated with hydroxycarbamide and showed resistance to this drug or its intolerance.

The assessment of the response was performed after 12 months of treatment. Overall response rate to interferon was higher in patients diagnosed with ET than in patients with polycythemia vera. In essential thrombocythemia, the percentage of achieved complete remissions was 43 and 26% of partial remissions. The remission rate in ET patients was higher if calreticulin CALR gene mutation was present. Patients with polycythemia vera achieved complete remission in 22% of cases and partial remission in 38% of cases.

Treatment-related side effects that follow to discontinuation of treatment were reported in almost 14% of patients [15].

The duration of response to treatment with pegylated interferon alpha-2a and the assessment of its safety in long-term use in patients with chronic myeloproliferative disorders was the goal of a phase II of the single-center study. Forty-three adult patients with polycythemia vera and 40 patients with essential thrombocythemia were enrolled in the study. The complete hematological response was defined as a decrease in hemoglobin concentration below 15.0 g/l, without phlebotomies, a resolution of splenomegaly, and no thrombotic episodes in the case of PV, and for essential thrombocythemia—a decrease platelet count below 440,000/μl and two other conditions as above. The assessment of the hematological response was performed every 3–6 months. The median follow-up was 83 months.

The hematological response was obtained in 80% of cases for the entire group. In patients with polycythemia vera, 77% of patients achieved a complete response (CR) while 7% a partial response (PR). The duration of response averaged 65 months for CR and 35 months for PR. In the group of patients diagnosed with essential thrombocythemia, CR was achieved in 73% and PR in 3%. The durance of CR was 58 months and PR was 25 months.

The molecular response for the entire group was achieved in 63% of cases.

The overall analysis showed that the duration of hematological remission and its achievement with pegylated interferon alpha-2a treatment is not affected neither by baseline disease characteristics nor JAK2 allele burden and disease molecular status. There was also no effect on age, sex, or the presence of splenomegaly.

During the course of the study, 22% of patients discontinued the treatment, because of toxicity. Toxicity was the greatest at the beginning of treatment. The starting dose was 450 μg per week and was gradually tapered off.

Thus, on the basis of the above observations, the researchers established that pegylated interferon alpha-2a may give long-term hematological and molecular remissions [16].

The assessment of pegylated interferon alpha-2a in group of patients diagnosed with polycythemia vera only was performed. The evaluation was carried out on a group of 27 patients. Interferon decreased the JAK2 V617F allele burden in 89% of cases. In three patients who were JAK2 homozygous at baseline, after the interferon alpha-2a treatment wild-type of JAK2 reappeared. The reduction of the JAK2 allele burden was estimated from 49% to an average 27%, and additional in one patient the mutant JAK2 allele was not detectable after treatment. It can therefore be postulated that the action of pegylated interferon alpha-2a is directed to cells of the polycythemia vera clone [17].

In 2005, the results of treatment by pegylated interferon alpha-2b of 21 patients diagnosed with polycythemia vera and 21 patients diagnosed with essential thrombocythemia were published. In the case of polycythemia vera in 14 patients, PRV-1 gene mutation was initially detected. In 36% of cases, PRV-1 expression normalized after treatment with pegylated interferon alpha-2b. For the entire group of 42 patients, the remission assessment showed that complete remission was achieved in 69% cases after 6 months of treatment. However, only in 19 patients remission was still maintained 2 years after the start of the study. Pegylated interferon alpha-2b was equally effective in patients with PV and ET. The use and the type of prior therapy did not affect the achievement of remission [18].

Another study with enrolled only PV patients included 136 patients. They were divided into two arms. One group received interferon alpha-2b and the other group received hydroxycarbamide. Interferon dosage was administered in 3 million units three times a week for 2 years and then 5 million units two times a week. Hydroxycarbamide was administered at a dose between 15 and 20 mg/kg/day.

In the group of patients treated with interferon, a significantly lower percentage of patients developed erythromelalgia (9.4%) and distal parasthesia (14%) compared with the group treated with hydroxycarbamide, for whom these percentages were respectively: 29 and 37.5%. Interferon alpha-2b was found to be more effective in inducing a molecular response, which was achieved in 54.7% of cases, in comparison with hydroxycarbamide—19.4% of cases, despite the fact that the percentage of achieved general hematological responses did not differ between the groups and amounted about 70%. The 5-year progression free period in the interferon group was achieved in a higher percentage (66%) than in the hydroxycarbamide group (46.7%) [19].

2.1.1 Ropeginterferon (monopegylated interferon alfa-2b)

The most recent form of interferon approved by the *European Medicines Agency* (EMA) for the treatment of patients is ropeginterferon. It is human recombinant interferon alpha-2b. Ropeginterferon is a monopegylated form of interferon alpha-2b. Ropeginterferon is conjugated with a two-arm methoxypolyethylene glycol (mPEG).

Thanks to these changes to the structure of the molecule, it was possible to achieve a significant increase in its half-life. Ropeginterferon can be administered subcutaneously to patients every 14 days. The clinical trials conducted so far have assessed the ropeginterferon dose from 50 micrograms to a maximum dose of 500 microgams administered as standard every 2 weeks. The possible dose change in case of side effects includes not only the reduction of the drug dose itself, but also the extension of the interval between doses. The extension of the dosing interval up to 4 weeks was assessed.

Ropeginterforn was approved in 2019 by the EMA for the use in patients diagnosed with polycythemia vera without splenomegaly, as monotherapy.

Ropeginterferon, like the previous forms of interferons used in treatment, is contraindicated in patients with severe mental disorders, such as severe depression. It is also a contraindication in patients with noncompensatory standard treatment of disorders of the thyroid gland as well as severe forms of autoimmune diseases. The safety profile of ropeginterferon is similar to that of other forms of alpha interferons. The most common side effects are flu-like symptoms [20].

Ropeginterferon has been shown to exhibit in vitro activity against JAK2-mutant cells. The activity of ropeginterferon against JAK2-positive cells is similar to that of other forms of interferons used actually for standard therapy. Ropeginterferon has an inhibitory effect on erythroid progenitor cells with a mutant JAK2 gene. At the same time, it has almost no effect on progenitor cells without the mutated allele (JAK2-wile-type) and normal CD34+ cells. A gradual decrease of JAK2-positive cells was observed in patients with PV during ropeginterferon treatment. The examination was performed after 6 and 12 months of treatment. In comparison, the reduction in the percentage of JAK2 positive cells in patients treated with hydroxycarbamide was significantly lower.

These results may suggest that ropeginterferon may cause elimination of the mutant clone, but further prospective clinical trials are needed to confirm this theory. The evaluation was performed on a group of patients enrolled in the PROUD-PV study who were treated in France [21].

In 2017, a multicenter study was opened in Italy. The study was of the second phase. In total, 127 patients with polycythemia vera were included in the study. All patients enrolled on the study had low-risk PV. The clinical trial consisted of two arms. Patients received phlebotomies and low-dose aspirin in one arm and ropeginterferon in the other arm. The aim of the study was to achieve a hematocrit of 45% or lower without any evidence of disease progression. Ropeginterferon was administered every 2 weeks at a constant dose of 100 µg.

The response to the treatment was assessed after 12 months. The reduction of hematocrit to the assumed level was achieved in significantly higher percentage of patients in the ropeginterferon group than of patients who received only phlebotomies and aspirin. In addition, none of the patients treated with ropeginterferon experienced disease progression during the course of the study, while among those treated with phlebotomies, 8% of patients progressed.

Grade 4 or 5 adverse events were not observed in patients treated with ropeginterferon, and the incidence of remaining adverse event (AE) was small and comparable in both arms. The most common side effects in the ropeginterferon group were flu-like symptoms and neutropenia; however, the third-grade neutropenia was the most common (8% of cases) [22].

One of the most important clinical studies on the use of ropeginterferon was the PROUD-PV study and its continuation: the CONTINUATION-PV study. These were three-phase, multicenter studies. The aim of the study was to compare the effectiveness of ropeginterferon in relation to hydroxycarbamide. The study included adult patients diagnosed with polycythemia vera treated with hydroxycarbamide for less than 3 years and no cytoreductive treatment at all. In total, 257 patients received this treatment. The patients were divided into two groups: those receiving ropeginterferon or the other being given hydroxycarbamide.

During the PROUD-study, drug doses were increased until the hematocrit was achieved below 45% without the use of phlebotomies, and the normalization of the number of leukocytes and platelets was reached.

The PROUD-PV study lasted 12 months. After this time, the patients continued the treatment under the CONTINUATION-PV study for further 36 months. After the final analysis performed in the 12th month at the end of PROUD study, it was found that the hematological response rates did not differ between the ropeginterferon and hydroxycarbamide treatment groups. These were consecutively 43% in the ropeginterferon arm and 46% in the control arm.

However, after analyzing the CONTINUATION- PV study, it turned out that after 36 months of treatment, the rates of hematological responses begin to prevail in the group of patients receiving ropeginterferon, 53% versus 38% in the control group. Thus, from the above data, it can be seen that the response rate to ropeginterferon increases with the duration of treatment [23].

Another analysis of patients participating in the PROUD and CONTINUATION studies was based on the assessment of treatment results after 24 months, dividing patients into two groups according to age (under and over 60 years).

The initial comparison of both groups of patients showed that older patients had a more aggressive course of the disease. Patients over 60 years of age had a higher percentage of cells with a mutant JAK2 allele. They experienced both general symptoms and some complications, such as thrombosis, more frequently. Both patients under 60 years of age and over 60 years of age in the ropeginterferon arm had a higher rate of molecular response, namely 77.1 and 58.7% compared with the HU remission: 33.3 and 36.1%, respectively. Significantly higher reductions in the JAK2 allele were observed in both groups of patients after ropeginterferon treatment: it was 54.8% for younger patients and 35.1% for elderly patients. For comparison, this difference in the group of patients treated with HU was 4.5 and 18.4%, respectively.

What is more, the age did not affect the frequency of ropeginterferon side effects. In addition, the incidence of adverse ropeginterferon disorders was similar to that observed in the hydroxycarbamide group [24].

2.2 Essential thrombocythemia

Essential thrombocythemia is a clonal growth of multipotential stem cells in the bone marrow. The consequence of this is increased proliferation of megakaryocytes in the bone marrow and an increase in the number of platelets in the peripheral blood. The level of platelets above 450,000/μl is considered a diagnostic criterion.

Essential thrombocythemia may progress over time to a more aggressive form of myeloproliferation, i.e., myelofibrosis. The disease can also evolve into acute myeloid leukemia or myelodysplastic syndrome, both with very poor prognosis. Thromboembolic complications are serious, and they concern over 20% of patients. Thrombosis occurs in the artery and venous area. Moreover, in patients with a very high platelet count, above 1,000,000/μl, bleeding may occur as a result of secondary von Willebrand syndrome [1, 2].

The treatment of ET is primarily aimed to prevent thrombotic complications.

In low-risk patients, only acetylsalicylic acid is used. In cases of high-risk patients, hydroxycarbamide is the first-line drug for most patients. Anagrelide and interferon are commonly used as second-line drugs.

Due to the possible effects of hydroxycarbamide of cytogenetic changes in the bone marrow cells after long-lasting usage, some experts recommend the use of interferon in younger patients in the first line. Interferon is also used as the drug of choice in patients planning a pregnancy [25].

The efficacy of pegylated interferon alpha-2a was assessed on the basis of the group of 39 patients with essential thrombocythemia and 40 patients with polycythemia vera.

Of the overall group, 81% of patients were previously treated prior to the study entry. The patients received pegylated interferon alpha-2a in a dose of 90 μg once a week. The dose of 450 μg was associated with a high percentage of intolerance.

In patients with essential thrombocythemia, the complete remission was achieved in 76%, while the overall hematological response rate brought 81%. Moreover, the molecular remission was achieved in 38%, in 14% of cases, JAK2 transcript became not detectable.

Patients diagnosed with polycythemia vera achieved 70% complete hematological remission and 80% general hematological response to treatment. JAK2 transcript was undetectable in 6% of patients. Molecular remission was achieved in 54% of cases.

Pegylated interferon alpha-2a at the dose of 90 μg per week was very well tolerated. In total, 20% of patients experienced a grade of 3 or 4 of adverse reaction, which was neutropenia. In addition, an increase in liver function tests was observed. Grade 4 of AE was not observed among patients who started the treatment with 90 μg/week while grade 3 neutropenia was an adverse event in only 7% of cases [26].

The effect of interferon alpha-2b treatment in patients with ET and PV was investigated. The study was prospective. Some of the results concerning the group of patients with polycythemia vera are presented in the subsection on polycythemia vera. In total, 123 patients with diagnosed essential thrombocythemia participated in the study. All of them received interferon alpha-2b. The patients were divided into two groups depending on the presence of the JAK2 V617F mutation. The enrolled patients were between 18 and 65 years of age. The treatment they received was, sequentially, interferon alpha-2b in the dose of 3 million units three times a week for the first 2 years, after which time the dose was changed into a maintenance dose, which amounted to 5 million units two times a week.

The analysis showed that the patients with the JAK2 V617F mutation present in a higher percentage achieved an overall hematological response as well as a complete hematological response. The overall hematological response was achieved in 83% of patients with JAK2 mutation, and the complete hematological remission was achieved in 23 cases. In the group of ET patients without the JAK2 V617F mutation, overall hematological response was achieved in 61.4%, while the complete hematological remission was achieved in 12 patients. The 5-year progression-free survival was obtained in 75.9% in the JAKV617F group and only in 47.6% without the mutation.

A significant proportion of patients experienced mild side effects. Grade 3 and 4 of adverse events were severe, most of them being a fever. The isolated cases of elevated liver tests and nausea have also been reported [19].

Pegylated interferon alpha-2b in patients with essential thrombocythemia who were previously treated with hydroxycarbamide, anagrelide, and other forms of interferon alpha, however, due to the lack of efficacy or toxicity, the patients required a change of treatment, was assessed. Pegylated interferon alpha-2b turned out to be effective in these cases. It led to the complete hematological remission in 91% of patients after 2 months of therapy, and in 100% of patients after 4 months. However, merely 11 patients participated in the study. Also only two patients required treatment discontinuation due to the side effects such as depression and general fatigue grade 3 [27].

2.2.1 Pregnancy

In case of pregnant patients, interferon is currently considered the only safe cytoreductive drug. Over the years, several analyses of the results of interferon treatment during pregnancy have been carried out.

The assessment of 34 pregnancies in 23 women diagnosed with ET was performed retrospectively. All the pregnancies included in the analysis were of high risk. This high risk was associated with a high platelet count above 1,500,000/μl, a history of thrombotic episode, severe microcirculation disorders, or a history of major hemorrhage.

It turned out that the use of interferon allowed the birth of an alive child in 73.5% of cases. There was no difference in efficacy between the basic and pegylated forms of interferon alpha. In pregnancies without interferon treatment, the percentage of live births was only 60%. Moreover, it was not found if the presence of the JAK2 V617F mutation had any influence on the course of pregnancy [28].

An analysis of the course of pregnancy in patients with ET was assessed in Italy. Data from 17 centers were taken into account. Data from 122 pregnancies were collected from 92 women. In patients diagnosed with essential thrombocythemia, the risk of the spontaneous loss of pregnancy is about 2.5 times higher than among the general population. In the contrary to the study quoted above, it was found that the presence of the JAK2 mutation increases the risk of pregnancy loss. The proportion of live births in patients exposed to interferon during pregnancy was 95%, compared with 71.6% in the group of patients not treated with interferon.

The multivariate analysis also showed that the use of acetylsalicylic acid during pregnancy had no effect on the live birth rate of patients with ET [29].

Whatever its form, interferon is the drug of first choice in pregnancy. Hydroxycarbamide and anagrelide should be withdrawn for about 6 months, and at least for 3 months, before the planned conception. Experts recommend the use of interferon in high-risk pregnancies [30]. A Japanese analysis of 10 consecutive pregnancies in ET patients showed 100% live births in patients who received interferon [31].

2.3 Myelofibrosis

In myelofibrosis (MF), monoclonal megakaryocytes produce cytokines that stimulate the proliferation of normal, non-neoplastic fibroblasts and stimulate angiogenesis. The consequence of this is the gradual fibrosis of the bone marrow, impaired hematopoiesis in the bone marrow, and the formation of extramedullary location mainly in the sites of fetal hematopoiesis, i.e., in the spleen and the liver.

The production of various cytokines by neoplastic megakaryocytes leads to the proliferation of normal, noncancerous fibroblasts as well as to increased angiogenesis.

Progressive bone marrow fibrosis leads to worsening anemia and thrombocytopenia. On the other hand, the production of proinflammatory cytokines by megakaryoblasts leads to the general symptoms such as weight loss, fever, joint pain, night sweats, and consequently, progressive worsening of general condition.

The prognosis for myelofibrosis is poor. In about 20% of patients, myelofibrosis evolves into acute myeloid leukemia with poor prognosis.

Currently, the only effective method of treatment that gives a chance to prolong the life is allogeneic bone marrow transplantation. However, this method is only available to younger patients.

The goal of treatment of patients who have not been qualified for allotranspalntation is to reduce the symptoms and to improve the patient's quality of life. In case of leukocytosis cytoreducing drugs, such as hydroxycarbamide, melphalan, or cladribine can be used. They cause a reduction in the number of leukocytes and may, to some extent, inhibit splenomegaly. Interferon alpha has been used successfully for the treatment of myelofibrosis for many years. The results of its effectiveness will be presented below [2].

Currently, the JAK2 inhibitor ruxolitinib is approved for the treatment of myelofibrosis with enlarged spleen in intermediate and high-risk patients. Ruxolitinib reduces the size of the spleen, reduces general symptoms, and improves the quality of life; however, it does not prolong the overall survival of patients [32].

In 2015, the results of a retrospective study were published to compare the histological parameters of the bone marrow before and after interferon treatment. Twelve patients diagnosed with primary myelofibrosis as well as post-PV MF and post-ET MF were enrolled in the study. Patients were treated with pegylated recombinant interferon alpha-2a or recombinant interferon alpha-2b in standard doses. The time of treatment was from 1 to 10 years. Some patients had previously been treated with hydroxycarbamide or anagrelide. In all cases, karyotype was normal. The prognostic factor of Dynamic International Prognostic Scoring System (DIPSS) was assessed at the beginning as well as during the treatment.

Bone marrow cellularity decreased in cases with increased bone marrow cellularity before the treatment. After the interferon treatment, a reduction in the degree of bone marrow fibrosis was found. The parameters, such as the density of naked nuclei and the density of megakaryocytes in the bone marrow, also improved.

It proves that if the JAK2 V617F mutation had been present, DIPSS was decreased after interferon treatment. This relationship was not observed in patients without the JAK2 V617F mutation. The improvement in peripheral blood morphological parameters and the overall clinical improvement correlated with the improvement in the assessed histological parameters of the bone marrow.

Before the initiation of interferon, seven patients had splenomegaly. During the treatment with interferon, the complete resolution of splenomegaly was achieved in 17% of patients (two cases), and its size decreased in 25% (three cases). A good clinical response was achieved in 83% during interferon therapy. There was no significant difference in response between the two types of interferon used [33].

A prospective study was also conducted in patients with low and intermediate-1 risk group myelofibrosis. Seventeen patients were enrolled. Patients received interferon alpha-2b (0.5–3 milion units/three times a week) or pegylated interferon alpha-2a (45–90 μg/week). The duration of therapy was on average 3.3 years.

Most of the patients responded to the treatment. Partial remission was found in seven patients and complete remission in two patients. Moreover, in four cases, the disease was stabilized and in one case the clinical improvement was achieved. Three patients did not respond to treatment at all and progressed to myelofibrosis. Additionally, the assessment in reducing spleen size was performed. At baseline, 15 patients have splenomegaly, nine of them achieved the compete regression of spleen size [34].

However, the efficacy of interferon in the treatment of myelofibrosis appears to be limited only to a less advanced form, when the bone marrow still has an adequate percentage of normal hemopoiesis and the marrow stroma is not significantly fibrotic. In more advanced stages, interferon was not shown to have any significant effect on the regression of the fibrosis process [35].

In 2020, the results of the COMBI study were published. That was a two-phase, multicenter, single-arm study that investigated the efficacy and safety of the combination of ruxolitinib and pegylated interferon alpha. Thirty-two patients with PV and 18 patients with primary and secondary myelofibrosis participated in the study. The patients were at age 18 and older. Remission was achieved in 44% of myelofibrosis cases, including 28% (5 patients) of complete remission. In patients with PV, the results were slightly worse: 31% of remissions, including 9% of complete remissions. Patients received pegylated interferon alpha-2a (45 μg/week) or pegylated interferon alpha-2b (35 μg/week) in low doses and ruxolitinib in doses of 5–20 mg twice a day.

For the entire group of patients (with PV and MF), the initial JAK2 allele burden was 47% at baseline, and after 2 years of treatment with interferon and ruxolitinib, it decreased to 12%.

The treatment toxicity was low. The highest incidence of side effects occurred at initiation of therapy. It was mostly anemia and thrombocytopenia.

The observations from the COMBI study show that, for the combination of interferon in lower doses with ruxolitinib, it may be effective and well tolerated even in the group of patients who had intolerance to interferon used as the only drug in higher doses. The combined treatment improved the bone marrow in terms of fibrosis and its cellularity. It also allowed to improve the value of peripheral blood counts [36].

It is currently known that some of the additional mutations are associated with a worse prognosis in patients with myelorpoliferation, including patients with myelofibrosis. Some of these mutations have been identified as high-risk molecular mutations. These are ASXL1, EZH2, IDH1/2, or SRSF2. Earlier studies have shown their association with a more aggressive course of the disease, worse prognosis, and shorter survival of patients, as well as a poorer response to treatment. Due to their importance, they have been included in the diagnostic criteria of myelofibrosis [37].

It is also known that the presence of driver mutations, i.e., JAK2, CALR, and MPL or triple negativity, may affect the course of myeloproliferation, including the incidence of thromboembolic complications.

The assessment of the influence of driver mutations and a panel of selected additional mutations on the effectiveness of interferon treatment in patients with myelofibrosis was performed on a group of 30 patients. Only the patients with low- and intermediate-1-risk were enrolled in the study. The treatment with pegylated interferon alpha-2a or interferon alpha-2b resulted in a complete remission in two patients and partial remission in nine patients. The disease progressed in three cases. One patient relapsed and four died. The remaining patients achieved a clinical improvement or disease stabilization. In the studied group, it was not found if the effectiveness of interferon treatment was influenced by the lack of driver mutations. Among the group of four patients with additional mutations, two died and one had disease progression. It was a mutation of ASXL1 and SRSF2. The treatment with interferon in patients without additional molecular mutations in the early stages of the disease may prevent further progression of the disease [38].

The side effects of interferon in the group of patients with myelofibrosis are similar to those occurring after the treatment of other chronic myeloproliferative diseases. The most frequently described are hematological toxicity- anemia and thrombocytopenia, less often is the appearance of leukopenia. Hematological toxicity usually resolves with dose reduction or extension of the dose interval. The most frequently nonhematological toxicity was fatigue, muscle pain, weakness, and depression symptoms. All symptoms are usually mild and do not exceed grade 2 [38].

However, the use of interferon in the treatment of myelofibrosis has not been recommended as a standard therapy. Interferon is still being evaluated in clinical trials, or it is used in selected patients as a nonstandard therapy in this diagnosis.

2.4 Mastocytosis

Mastocytosis is characterized by an excessive proliferation of abnormal mast cells and their accumulation in various organs.

The basis for the development of mastocytosis is ligand-independent activation of the KIT receptor, resulting from mutations in the KIT proto-oncogene. The KIT receptor is a trans membrane receptor with tyrosine kinase's activity. Its activation stimulates the proliferation of mast cells. That excessive numbers of mast cells infiltrate tissues and organs and release mediators such as histamine, interleukine-6, tryptase, heparin, and others, which are responsible for the appearance of symptoms typical of mastocytosis. In addition, the infiltration of tissues for mast cells itself causes damage to the affected organs.

The prognosis of mastocytosis depends on the type of the disease. In the case of cutaneous mastocytosis (CM), in the majority of cases prognosis is good and the disease does not shorten the patient's life, but in aggressive systemic mastocytosis

(ASM), the average follow-up is about 40 months. Mast cell leukemia has a poor prognosis with a median follow-up of approximately 1 year.

Systemic mastocytosis usually requires the implementation of cytoreductive therapy. The first line of therapy is interferon alone or its combination with corticosteroids. In aggressive systemic mastocytosis, the first line in addition to interferon 2-CdA can be used. An effective drug turned out to be midostaurin in the case of the present KIT mutation. In patients without the KIT D816V mutation, treatment with imatinib may be effective. In the case of mast cell leukemia, multidrug chemotherapy is most often required, as in acute leukemias, followed by bone marrow transplantation [39].

Systemic mastocytosis requiring treatment is a rare disease, this is why the studies available in the literature evaluating various therapies concern mostly small groups of patients.

In 2002, the French authors presented their experiences on the use of interferon in patients with systemic mastocytosis. They included 20 patients. The patients received interferon alpha-2b in gradually increased doses.

The patients were assessed after 6 months. In cases in which bone marrow was infiltrated for mast cells at baseline, it still remained infiltrated after 6 months of treatment.

However, the responses were obtained in terms of symptoms related to mast cell degranulation. Partial remission was achieved in 35% of patients and minor remission in 30%. It concerns mainly skin lesions and vascular congestion. Moreover, the assessment of the histamine level in the plasma revealed a decrease of it in patients who previously presented symptoms related to the degranulation of mast cells, such as gastrointestinal disorders and flushing.

A high percentage of side effects were found during treatment. They concerned 35% of patients. Depression and cytopenia were most frequent ones [40].

Another analysis was a report of five patients with systemic mastocytosis treated with interferon and prednisolone. All patients received interferon alpha-2b in a dose of 3 million units three times a week and four patients additionally received prednisolone. Four patients responded to interferon treatment at varying degrees. One patient, who at baseline had bone marrow involvement by mast cells in above 10%, progressed to mast cell leukemia. In two patients, the symptoms C resolved completely and in one of them they partially disappeared. In one case, stabilizing disease was achieved [41].

In 2009, a retrospective analysis of patients treated with cytoreductive therapy due to mastocytosis was published. The authors collected data from 108 patients treated at the Mayo Clinic. This analysis allowed for the comparison of the efficacy of four drugs used in systemic mastocytosis. There were interferon alpha alone or in the combination with prednisone—among 40 patients, hydroxycarbamide—among 26 ones, imatinib—among 22 persons, and 2-chlorodeoxyadenosine (2-CdA)—among 22 patients.

After dividing the patients into three additional groups on the basis of the type of mastocytosis—indolent systemic mastocytosis, aggressive systemic mastocytosis, and systemic mastocytosis associated with another clonal hematological nonmast cell lineage disease (SM-AHNMD)—the effectiveness of each of type of therapy was assessed.

The highest response rates in indolent and aggressive mastocytosis were achieved with interferon treatment. They were 60% of the responses in both groups, and in the SM-AHNMD group of patients, the percentage was also one of the highest and amounted to 45%. The second most effective drug was 2-CdA. The response rates were 56% for indolent MS, 50% for aggressive MS, and 55% for SM-AHNMD. The patients treated with imatinib achieved response in 14, 50, and 9% by following groups, respectively. In contrast, patients with indolent and aggressive systemic mastocytosis did not respond to hydroxycarbamide treatment at all. The response rate in both groups was 0%. However, patients with MS associated with another clonal hematological nonmast cell lineage disease achieved 21% response to hydroxycarbamide. Additionally, it was found that only interferon relieved symptoms caused by the release of inflammatory mediators by mast cells.

The additional analysis showed no influence of the TET 2 mutation on the response to treatment [42].

In the literature, there are also single cases of mastocytosis presenting trials of non-standard treatment. That is description of a patient with systemic mastocytosis with mast cell bone marrow involvement. Mutation of c-kit Asp816Val was present. Patient progressed despite treatment with dasatinib and 2-chlorodeoxyadenosine. The patient developed symptoms related to the degranulation of mast cells and increased ascites.

The patient was treated with pranlukast, which is an anti-leukotriene receptor antagonist due to an asthma episode. The rate of ascites growth decreased significantly after one administration. The patient required paracentesis every 10 days and not every 3 days, as before starting to take the drug. After 15 days of treatment with pranlukast, the patient received interferon alpha, which resulted in complete regression of ascites, resolution of pancytopenia, and complete disappearance of the c-kit mutation clone. The infiltration of mast cells in the bone marrow significantly decreased [43].

Interferon alpha was also effective in a patient with systemic mastocytosis associated with myelodysplastic syndrome with the c-kit D816V mutation, which was refractory to imatinib treatment [44].

Interferon alpha also proved to be effective in the treatment of osteoporotic lesions appearing in the course of mastocytosis.

The series of 10 cases with resolved mastocytosis and osteoporosis-related fractures was presented in 2011. The patients received interferon alpha in a dose of 1.5 million units three times a week as well as pamindronic acid. The patients were treated for an average of 60 months. For the first 2 years, pamindronate was given at a dose of 1 mg/kg every month, and then every 3 months.

During the course of the study, no patient had a new-bone fracture. The level of alkaline phosphatase decreased by 25% in relation to the value before treatment and tryptase by 34%. Bone density increased during treated with interferon and pamindronate. The increase was on average 12% in the spine bones and 1.9% in the hip bones. At the same time, there was no increase in the density of the hip bone and a minimal increase in the density of the spine in patients treated with pamindronate alone.

The results of this observation suggest that it is beneficial to add low doses of interferon alpha to pamindronate treatment in terms of bone density increase [45].

That experiences show that interferon used in systemic mastocytosis significantly improves the quality of life of patients by inhibiting the symptoms caused by degranulation of mast cells. They prevent bone fractures and, in some patients, they cause remission of bone marrow infiltration by mast cells.

2.5 Chronic neutrophilic leukemia

Chronic neutrophilic leukemia (CNL) is a very rare disease. It is characterized by the clonal proliferation of mature neutrophils.

The diagnostic criteria proposed by the World Health Organization (WHO) comprise leukocyte counts above 25,000/µl (including more than 80% of rod and segmented *neutrophils* in the bone marrow blast cells count below 5%), normal neutrophils maturation, and an increase of neutrophilopoiesis. Also the presence of the CSF3R gene mutation is required.

Physical examination often shows enlargement of the liver and spleen, moreover, patients complain on weight loss and weakness [1].

The prognosis varies. The average survival time for patients with CNL is less than 2 years.

Only few descriptions of chronic neutrophilic leukemia are available in the literature, and these are mostly single case reports.

Because it is an extremely rare disease, there are no established and generally accepted treatment standards. In most cases, patients are given hydroxycarbamide or interferon. Patients who are eligible for a bone marrow transplant may benefit from this treatment. Bone marrow allotransplantation remains the only method that gives a chance for a significant extension of life.

The German authors presented a series of 14 cases of chronic neutrophilic leukemia. The group of patients consisted of eight women and six men. The average age was 64.7 years. From the entire group of patients, longer survival was achieved only in three cases. One of these patients was treated with interferon alpha and achieved hematological remission, the other underwent bone marrow allotransplantation from a family donor, and the third one was treated with hydroxycarbamide and transfusions as needed. The follow-up period of the patient after allogeneic matched related donor transplantation (allo-MRD) was 73 months, and for the patient after interferon treatment it was 41 months.

The remaining patients died within 2 years of diagnosis. Six patients, the largest group, died due to intracranial bleeding, three patients died because of leukemia cell tissue infiltration, one patient because of the disease transformation into leukemia, and one patient because of pneumonia [46].

It can be seen from these experiences that treatment with interferon alpha can significantly extend the survival time of patients.

The case of a 40-year-old woman diagnosed with chronic neutrophilic leukemia is presented by Yassin and coauthors. Initially, the patient had almost 41,000 leukocytes in the peripheral blood. In a physical examination, splenomegaly and hepatomegaly were not present. Patient received pegylated interferon alpha-2a. The initially dose was 50 μg once a week for the first 2 weeks, then the dose was increased to 135 μg weekly for 6 weeks, and then the dose interval was extended to another 2 weeks. As a result of the treatment, the general condition of the patient improved and the parameters of peripheral blood counts were normalized [47].

Another case report presented in the literature describes a 41-year-old woman diagnosed with CNL accompanied by focal segmental glomerulosclerosis (FSGS). The patient had increasing leukocytosis for several months. On the admission to the hospital, leukocytosis was 94,000/μl. Moreover, the number of platelets in the morphology exceeded 1,000,000/μl. More than a year earlier, the patient had splenectomy due to splenomegaly and spleen infraction.

Additionally, JAK2 V617F mutation was found. Some authors suggest that the presence of JAK2 mutation may be associated with longer survival in CNL.

The patient received hydroxycarbamide for 3 months and reduction in the number of leukocytes was achieved. After this time, interferon alpha-2b was added to hydroxycarbamide. As a result, focal segmental glomerulosclerosis disappeared and the renal tests improved [48].

Another case of chronic neutrophilic leukemia with a JAK2 gene mutation concerns a 53-year-old man. The patient's baseline leukocytosis was 33,500/µl, including the neutrophil count of 29,700/µl. The patient also had splenomegaly.

The treatment with interferon alpha-2b at a dose of 3 million units every other day was started. After a month of treatment, the number of leukocytes was reduced to less than 10,000/µl. Then the patient was treated chronically with interferon alpha-2b in doses of 3 million units every 2 weeks. As a result of the therapy, the number of leukocytes remains between 8 and 10,000/µl. The patient remains in general good condition [49].

A series of two CNL cases are also shown. The first patient was a 70-year-old woman with stable leukocytosis of about 35,000/µl and the remaining morphology parameters in normal range. The patient was only observed for 5 years until hepasplenomegaly progressed rapidly. Then, interferon alpha-2b was included. Due to the treatment, the rapid regression of hepatosplenomegaly was achieved.

The second case is a 68-year-old woman with baseline leukocytosis of almost 14,000/µl. In this case, the treatment with hydroxycarbamide was started immediately. However, no improvement was achieved. After 6 weeks of HU treatment, interferon alpha-2b 3 million units 3 times a week was implemented and leukocytosis decreased. Due to the interferon treatment, the disease stabilized for a long time. Because the patient experienced an adverse reaction, a severe flu-like syndrome, interferon was discontinued. After interferon withdrawal, the disease progressed gradually and the treatment attempts by busulfan and 6-mercaptopurine were unsuccessful. Therefore, interferon was readministered and the disease went into remission. Interferon treatment was continued at a reduced dose. The disease regression was achieved again.

Additionally, the patient showed an improvement in the function of granulocytes in terms of phagocytosis and an improvement in neutral killer (NK) cell function after treatment with interferon [50].

The above examples show that interferon alpha is effective in the treatment of chronic neutrophilic leukemia. The side effects are rare and can be managed with dose reductions. Moreover, in these cases, interferon is also effective in a reduced dose. Disease remission or regression can be achieved without typical of CNL complications, such as intracranial bleeding.

2.6 Another

Interferon has been used in the past to treat chronic myeloid leukemia. The treatment with tyrosine kinase inhibitors is now a standard practice. However, in a small number of patients, they are ineffective or exhibit unmanageable toxicity. Therefore, the attempts are underway to use interferon in combination with TKI in lower doses, which is to ensure the enhancement of the antiproliferative effect while reducing the toxicity.

There are ongoing attempts to use ropeginterferon in patients diagnosed with chronic myeloid leukemia, in whom treatment with imatinib alone has not led to deep molecular response (DMR). The first phase study was conducted in a small group of patients with chronic myeloid leukemia. The patients in first chronic phase treated with imatinib who did not achieve DMR, but in complete hematologic remission and complete cytogenetic remission, were included in the study. Patients have been treated with imatinib for at least 18 months. Twelve patients were enrolled in the study, and they completed the study according to the protocol. These patients

Source	Type of trial	Interferon	Diagnosis	No.	Prior treatment status	Response rate
Yacoubet al. [15]	Phase II, multicenter	Pegylated IFN alfa-2a	PV	50	Resistance to HU or HU intolerance	CR:22% PR:38%
			ET	65		CR:43% PR:26%
Masarova et al. [16]	Phase II, single-center	Pegylated IFN alfa-2a	PV	43	Untreated or previously treated with cytoreductive therapy	CR:77% PR:7%
			ET	40		CR:73% PR:3%
Samuelsson et al. [18]	Phase II	Pegylated IFN alfa-2b	PV	21	Untreated or previously treated with cytoreductive therapy	CR: 69% for the entire group
			ET	21		
Huang BT et al. [19]	Open label, multicenter	IFN alfa-2b	PV	136	Untreated or previously treated with cytoreductive therapy	OHR:70% Molecular response:54.7%
			ET	123		OHR (JAK2+ patients):83% CHR:23 cases OHR (JAK2-patients): 61.4% CHR:12 cases
Gisslinger et al. [23]	phase III, multicenter	Ropeginterferon	PV	257	Previously treated	OHR:53%

DOI: http://dx.doi.org/10.5772/intechopen.104501

Source	Type of trial	Interferon	Diagnosis	No.	Prior treatment status	Response rate
Quintás-Cardama et al. [26]	phase II	Pegylated IFN alfa-2a	PV	40	Untreated or previously treated with cytoreductive therapy	OHR:80% CR:70% Molecular remission:54%
			ET	39		OHR:81% CR:76% Molecular remission:38%
Sørensen et al. [36]	Phase III, multicenter, COMBI	Pegylated IFN alfa-2a with ruxolitinib or Pegylated IFN alfa-2b with ruxolitinib	PV	32	Untreated or previously treated with cytoreductive therapy	OHR:44% CR:28%
			MF	18		OHR:31% CR:9%
Casassus et al. [40]	Open label, multicenter	IFN alpha-2b	Mastocytosis	20	Untreated and previously treated	PR:35% Minor remission: 30%

PV: polycythemia vera; ET: essential thrombocythemia; MF: myelofibrosis; HU: hydroxycarbamide/hydroxyurea; CR: complete remission; PR: partial remission; and OHR: overall hematological response.

Table 1.
Comparison of the effectiveness of interferon in chronic myeloproliferative disorders.

received additional ropeginterferon to imatinib and four achieved DMR. Low toxicity was observed during the treatment. Among the hematological toxicities, neutropenia was the most common. There was no nonhematological toxicity with a degree higher than 1/2 during the treatment. Moreover, it has been found that better effects and fewer side effects are obtained when ropeginterferon is administered for a longer time, but in lower doses. The comparison of the effectiveness of interferon in chronic myeloproliferative disorders based on selected articles is presented in **Table 1** [51].

3. Conclusions

Interferon alpha appears to be an effective and safe drug in the most type of chronic myeloproliferative disorders. Nowadays, all forms of its using have similar effectiveness. Interferon alpha can be effective even in cases of resistance for first-line treatment. Trial research is currently underway to combine it with some new drugs, such as ruxolitinib, and to add it to the already well-established therapy, it is a promising option for patients with refractory disease.

From time to time, new forms of interferon, such as ropeginterferon, are introduced, which gives hope for better effectiveness, better safety profile, and greater comfort in its use for patients who have to be treated for many years. In the case of the use of interferons alpha in the treatment of chronic myeloproliferative diseases, there are still opportunities to extend its use and to study its combination with newly introduced drugs.

Author details

Anna Prochwicz[1*] and Dorota Krochmalczyk[2]

1 Department of Hematology, Specialized Hospital, Nowy Sącz, Poland

2 Department of Hematology, Jagiellonian University Medical College, Cracow, Poland

*Address all correspondence to: annaprochwicz@interia.pl

References

[1] Arber DA, Orazi A, Hasserjian R, et al. The 2016 revision to the World Health Organization classification of myeloid neoplasms and acute leukemia. Blood. 2016;**127**:2391-2405. DOI: 10.1182/blood-2016-03-643544

[2] Spivak JL, Barosi G, Tognoni G, et al. Chronic myeloproliferative disorders. Hematology. American Society of Hematology. Education Program. 2003:200-224. DOI: 10.1182/asheducation-2003.1.200

[3] Isaacs A, Lindenmann J. Virus interference. I. The interferon. Proceedings of the Royal Society of London—Series B: Biological Sciences. 1957;**147**:258-267. DOI: 10.1098/rspb.1957.0048

[4] Hasselbalch HC, Holmström MO. Perspectives on interferon-alpha in the treatment of polycythemia vera and related myeloproliferative neoplasms: Minimal residual disease and cure? Seminars in Immunopathology. 2019;**41**:5-19. DOI: 10.1007/s00281-018-0700-2

[5] Ludwig H, Linkesch W, Gisslinger H, et al. Interferon-alfa corrects thrombocytosis in patients with myeloproliferative disorders. Cancer Immunology, Immunotherapy. 1987;**25**:266-273. DOI: 10.1007/BF00199157

[6] Lu M, Zhang W, Li Y, et al. Interferon-alpha targets JAK2V617F-positive hematopoietic progenitor cells and acts through the p38 MAPK pathway. Experimental Hematology. 2010;**38**:472-480. DOI: 10.1016/j.exphem.2010.03.005

[7] Kiladjian JJ, Chomienne C, Fenaux P. Interferon-alpha therapy in bcr-abl-negative myeloproliferative neoplasms. Leukemia. 2008;**22**:1990-1998. DOI: 10.1038/leu.2008.280

[8] Wang Q, Miyakawa Y, Fox N, Kaushansky K. Interferon-alpha directly represses megakaryopoiesis by inhibiting thrombopoietin-induced signaling through induction of SOCS-1. Blood. 2000;**96**:2093-2099

[9] Waters MJ, Brooks AJ. JAK2 activation by growth hormone and other cytokines. The Biochemical Journal. 2015;**466**:1-11. DOI: 10.1042/BJ20141293

[10] Brooks AJ, Dai W, O'Mara ML, et al. Mechanism of activation of protein kinase JAK2 by the growth hormone receptor. Science. 2014;**344**:1249783. DOI: 10.1126/science.1249783

[11] Chaligné R, Tonetti C, Besancenot R, et al. New mutations of MPL in primitive myelofibrosis: Only the MPL W515 mutations promote a G1/S-phase transition. Leukemia. 2008;**22**:1557-1566. DOI: 10.1038/leu.2008.137

[12] Chachoua I, Pecquet C, El-Khoury M, et al. Thrombopoietin receptor activation by myeloproliferative neoplasm associated calreticulin mutants. Blood. 2016;**127**:1325-1335. DOI: 10.1182/blood-2015-11-681932

[13] Kralovics R, Passamonti F, Buser AS, et al. A gain-of-function mutation of JAK2 in myeloproliferative disorders. The New England Journal of Medicine. 2005;**352**:1779-1790. DOI: 10.1056/NEJMoa051113

[14] Hino M, Futami E, Okuno S, et al. Possible selective effects of interferon alpha-2b on a malignant clone in a case of polycythemia vera. Annals of

Hematology. 1993;**66**:161-162. DOI: 10.1007/BF01697629

[15] Yacoub A, Mascarenhas J, Kosiorek H, et al. Pegylated interferon alfa-2a for polycythemia vera or essential thrombocythemia resistant or intolerant to hydroxyurea. Blood. 2019;**134**:1498-1509. DOI: 10.1182/blood.2019000428

[16] Masarova L, Patel KP, Newberry KJ, et al. Pegylated interferon alfa-2a in patients with essential thrombocythaemia or polycythaemia vera: a post-hoc, median 83 month follow-up of an open-label, phase 2 trial. The Lancet Haematology. 2017;**4**:e165-e175. DOI: 10.1016/S2352-3026(17)30030-3

[17] Kiladjian JJ, Cassinat B, Turlure P, et al. High molecular response rate of polycythemia vera patients treated with pegylated interferon alpha-2a. Blood. 2006;**108**:2037-2040. DOI: 10.1182/blood-2006-03-009860

[18] Samuelsson J, Hasselbalch H, Bruserud O, et al. Nordic study Group for Myeloproliferative Disorders. A phase II trial of pegylated interferon alpha-2b therapy for polycythemia vera and essential thrombocythemia: Feasibility, clinical and biologic effects, and impact on quality of life. Cancer. 2006;**106**:2397-2405. DOI: 10.1002/cncr.21900

[19] Huang BT, Zeng QC, Zhao WH, et al. Interferon α-2b gains high sustained response therapy for advanced essential thrombocythemia and polycythemia vera with JAK2V617F positive mutation. Leukemia Research. 2014;**38**:1177-1183. DOI: 10.1016/j.leukres.2014.06.019

[20] European Medicines Agency. Besremi Assessment Report. European Medicines Agency Website. Amsterdam, Netherlands: European Medicines Agency; 2020. Available from: https://www.ema.europa.eu/en/medicines/human/EPAR/besremi

[21] Verger E, Soret-Dulphy J, Maslah N, et al. Ropeginterferon alpha-2b targets JAK2V617F-positive polycythemia vera cells in vitro and in vivo. Blood Cancer Journal. 2018;**8**:94. DOI: 10.1038/s41408-018-0133-0

[22] Barbui T, Vannucchi AM, De Stefano V, et al. Ropeginterferon alfa-2b versus phlebotomy in low-risk patients with polycythaemia vera (low-PV study): A multicentre, randomised phase 2 trial. The Lancet Haematology. 2021;**8**:e175-e184. DOI: 10.1016/S2352-3026(20)30373-2

[23] Gisslinger H, Klade C, Georgiev P, et al. PROUD-PV study group. Ropeginterferon alfa-2b versus standard therapy for polycythaemia vera (PROUD-PV and CONTINUATION-PV): A randomised, non-inferiority, phase 3 trial and its extension study. The Lancet Haematology. 2020;**7**:e196-e208. DOI: 10.1016/S2352-3026(19)30236-4

[24] Gisslinger H, Klade C, Georgiev P, et al. Ropeginterferon alfa-2b: Efficacy and safety in different age groups. Hema. 2020;**4**:e485. DOI: 10.1097/HS9.0000000000000485

[25] Geyer HL, Mesa RA. Therapy for myeloproliferative neoplasms: When, which agent, and how? Blood. 2014;**124**:3529-3537. DOI: 10.1182/blood-2014-05-577635

[26] Quintás-Cardama A, Kantarjian H, Manshouri T, et al. Pegylated interferon alfa-2a yields high rates of hematologic and molecular response in patients with advanced essential thrombocythemia and polycythemia vera. Journal of Clinical Oncology. 2009;**27**:5418-5524. DOI: 10.1200/JCO.2009.23.6075

[27] Alvarado Y, Cortes J, Verstovsek S, et al. Pilot study of pegylated interferon-alpha 2b in patients with essential thrombocythemia. Cancer Chemotherapy and Pharmacology. 2003;**51**:81-86. DOI: 10.1007/s00280-002-0533-4

[28] Schrickel L, Heidel FH, Sadjadian P, et al. German study group MPN, GSG-MPN. Interferon alpha for essential thrombocythemia during 34 high-risk pregnancies: Outcome and safety. Journal of Cancer Research and Clinical Oncology. 2021;**147**:1481-1491. DOI: 10.1007/s00432-020-03430-4

[29] Melillo L, Tieghi A, Candoni A, et al. Outcome of 122 pregnancies in essential thrombocythemia patients: A report from the Italian registry. American Journal of Hematology. 2009;**84**:636-640. DOI: 10.1002/ajh.21504

[30] Harrison CN, Bareford D, Butt N, et al. British Committee for Standards in Haematology. Guideline for investigation and management of adults and children presenting with a thrombocytosis. British Journal of Haematology. 2010;**149**:352-375. DOI: 10.1111/j.1365-2141.2010.08122.x

[31] Edahiro Y, Yasuda H, Gotoh A, et al. Interferon therapy for pregnant patients with essential thrombocythemia in Japan. International Journal of Hematology. 2021;**113**:106-111. DOI: 10.1007/s12185-020-03001-w

[32] Harrison CN, Vannucchi AM, Kiladjian JJ, et al. Long-term findings from COMFORT-II, a phase 3 study of ruxolitinib vs best available therapy for myelofibrosis. Leukemia. 2016;**30**:1701-1707. DOI: 10.1038/leu.2016.148

[33] Pizzi M, Silver RT, Barel A, Orazi A. Recombinant interferon-α in myelofibrosis reduces bone marrow fibrosis, improves its morphology and is associated with clinical response. Modern Pathology. 2015;**28**:1315-1323. DOI: 10.1038/modpathol.2015.93

[34] Silver RT, Vandris K, Goldman JJ. Recombinant interferon-alpha may retard progression of early primary myelofibrosis: A preliminary report. Blood. 2011;**117**:6669-6672. DOI: 10.1182/blood-2010-11-320069

[35] Bachleitner-Hofmann T, Gisslinger H. The role of interferon-alpha in the treatment of idiopathic myelofibrosis. Annals of Hematology. 1999;**78**:533-538. DOI: 10.1007/s002770050554

[36] Sørensen AL, Mikkelsen SU, Knudsen TA, et al. Ruxolitinib and interferon-α2 combination therapy for patients with polycythemia vera or myelofibrosis: A phase II study. Haematologica. 2020;**105**:2262-2272. DOI: 10.3324/haematol.2019.235648

[37] Vannucchi AM, Lasho TL, Guglielmelli P, et al. Mutations and prognosis in primary myelofibrosis. Leukemia. 2013;**27**:1861-1869. DOI: 10.1038/leu.2013.119

[38] Silver RT, Barel AC, Lascu E, et al. The effect of initial molecular profile on response to recombinant interferon-α (rIFNα) treatment in early myelofibrosis. Cancer. 2017;**123**:2680-2687. DOI: 10.1002/cncr.30679

[39] Pardanani A. Systemic mastocytosis in adults: 2021 update on diagnosis, risk stratification and management. American Journal of Hematology. 2021;**96**:508-525. DOI: 10.1002/ajh.26118

[40] Casassus P, Caillat-Vigneron N, Martin A, et al. Treatment of adult systemic mastocytosis with interferon-alpha: Results of a multicentre phase II

trial on 20 patients. British Journal of Haematology. 2002;**119**:1090-1097. DOI: 10.1046/j.1365-2141.2002.03944.x

[41] Hauswirth AW, Simonitsch-Klupp I, Uffmann M, et al. Response to therapy with interferon alpha-2b and prednisolone in aggressive systemic mastocytosis: Report of five cases and review of the literature. Leukemia Research. 2004;**28**:249-257. DOI: 10.1016/s0145-2126(03)00259-5

[42] Lim KH, Pardanani A, Butterfield JH, et al. Cytoreductive therapy in 108 adults with systemic mastocytosis: Outcome analysis and response prediction during treatment with interferon-alpha, hydroxyurea, imatinib mesylate or 2-chlorodeoxyadenosine. American Journal of Hematology. 2009;**84**:790-794. DOI: 10.1002/ajh.21561

[43] Mizuno I, Tominaga R, Gomyo H, et al. Systemic mastocytosis with refractory ascites successfully treated with interferon-α and a leukotriene receptor antagonist. Japanese. 2019;**60**:1555-1559. DOI: 10.11406/rinketsu.60.1555

[44] Yoshida C, Takeuchi M, Tsuchiyama J, Sadahira Y. Successful treatment of KIT D816V-positive, imatinib-resistant systemic mastocytosis with interferon-alpha. Internal Medicine. 2009;**48**:1973-1978. DOI: 10.2169/internalmedicine.48.2294

[45] Laroche M, Livideanu C, Paul C, Cantagrel A. Interferon alpha and pamidronate in osteoporosis with fracture secondary to mastocytosis. The American Journal of Medicine. 2011;**124**:776-778. DOI: 10.1016/j.amjmed.2011.02.038

[46] Böhm J, Schaefer HE. Chronic neutrophilic leukaemia: 14 new cases of an uncommon myeloproliferative disease. Journal of Clinical Pathology. 2002;**55**:862-864. DOI: 10.1136/jcp.55.11.862

[47] Yassin MA, Kohla S, Al-Sabbagh A, et al. A case of chronic neutrophilic leukemia successfully treated with pegylated interferon alpha-2a. Clinical Medicine Insights: Case Reports. 2015;**8**:33-36. DOI: 10.4137/CCRep.S22820

[48] Wang R, Tong H, Wang H, et al. Nephrotic syndrome related to chronic neutrophilic leukemia. Internal Medicine. 2014;**53**:2505-2509. DOI: 10.2169/internalmedicine.53.2162

[49] Zhang X, Pan J, Guo J. Presence of the JAK2 V617F mutation in a patient with chronic neutrophilic leukemia and effective response to interferon α-2b. Acta Haematologica. 2013;**130**:44-46. DOI: 10.1159/000345851

[50] Meyer S, Feremans W, Cantiniaux B, et al. Successful alpha-2b-interferon therapy for chronic neutrophilic leukemia. American Journal of Hematology. 1993;**43**:307-309. DOI: 10.1002/ajh.2830430416

[51] Heibl S, Buxhofer-Ausch V, Schmidt S, et al. A phase 1 study to evaluate the feasibility and efficacy of the addition of ropeginterferon alpha-2b to imatinib treatment in patients with chronic phase chronic myeloid leukemia (CML) not achieving a deep molecular response (molecular remission 4.5)-AGMT_CML 1. Hematological Oncology. 2020;**38**:792-798. DOI: 10.1002/hon.2786

Section 3

Interferon and Infectious Discease

Chapter 4

Perspective Chapter: Interferon-Gamma in Natural Defence and Prevention of Leprosy

Pragya Santra, Prama Ghosh, Soumyadeep Ghosh, Anwesha Behera, Oyendrilla Mitra, Ishanee Das Sharma, Diya Adhikary, Asesh Banerjee and Prabuddha Gupta

Abstract

Mycobacterium leprae causes leprosy. *M. leprae* enters the body through the upper respiratory tract where it interacts with host's cells. Interferon (IFN) is a class of cytokines in human body that are released in case of viral and intracellular pathogen infection and they activate the immune cells to eradicate those pathogens. IFN-γ (Type-II IFN) confers immunity against bacterial, viral, and protozoan diseases. Loss of function mutations in IFN-γ results in poor immunity towards mildly virulent mycobacterium. Upon *M. leprae* invasion, monocytes enter the site of infection and differentiates into macrophages. IFN-γ induces endothelial cells (EC) of the pathogenic micro-environment to cause monocyte differentiation into pro-inflammatory M1 macrophages for immediate antimicrobial activity. This differentiation is ceased in the absence of endothelial cells. M1 macrophages are clinically more active than anti-inflammatory M2 macrophages induced by resting EC. The former produced higher amounts of pro-inflammatory cytokines in response to the TLR2/1 ligand of *M. leprae*. The former also showed elevation of vitamin D-associated antimicrobial pathway genes, which are required to counter *M. leprae*. In addition, the former accumulates less oxidised LDL to prevent growth of *M. leprae*. Thus, advancement of IFN-γ research would help in the design of next-generation anti-leprosy therapeutics.

Keywords: leprosy, IFN-γ, TH1, TH2, *M. leprae*, cytokines, immunity, tuberculoid leprosy, lepromatous leprosy, cell-mediated immunity

1. Introduction

Leprosy (Hansen's disease) is a complex, chronic, granulomatous dermato-neurological disease caused by *Mycobacterium leprae*. Leprosy is characterised by acute

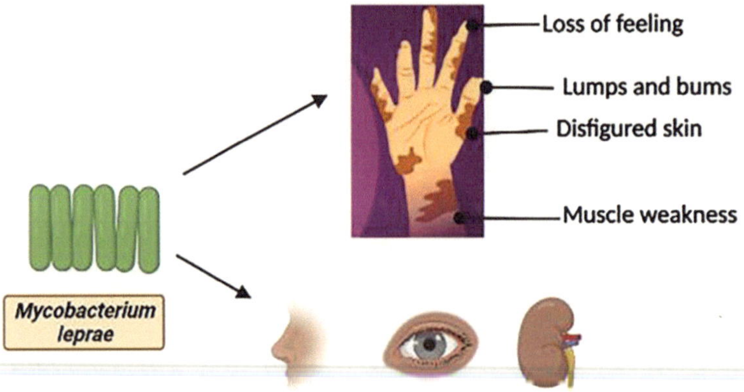

Figure 1.
Mycobacterium leprae *causes leprosy. In this disease, the skin and peripheral nerves are mainly affected. It also affects the nose, eyes, kidneys, etc.*

immunological reactions which lead to neural damage and disabilities with a myriad of clinical as well as serological manifestations. It includes the eyes, the mucosa of the upper respiratory tract, muscle, bone, and testes [1]. The infection by *Mycobacterium leprae* causes loss of sensation or numbness loss of feeling and lumps and bumps in the hands of the infected person along with disFig.d skin and muscle weakness; nose, eyes, and other internal organs, such as kidneys are also affected (**Figure 1**), thus characterised as broad-spectrum chronic infectious disease.

2. Transmission of leprosy

The pathogen is transmitted from an infected person to healthy individuals via aerosols harbouring the bacteria, especially infection by multibacillary patients supporting respiratory transmission [2]. The initial and common route for the pathogen is the upper respiratory tract, indicating that the interaction between the host and the bacteria initiates in the nasal passage [2]. The protective mucosal innate immune mechanism in the respiratory tract contributes to the low infectivity of the pathogen to some extent [3]. A detailed study of the nasal swab samples of patients by asymptomatic qPCR also indicates that the air route is a common entry canal for the bacteria. Thus, intimating that such contacts have a high chance of developing leprosy. The hypothesis of respiratory transmission is further validated by the adherence of *M. leprae* to alveolar and nasal epithelial cells [4].

3. Classifications of leprosy

Leprosy is not a single clinical entity but rather classified as a polymorphic infectious disease. The manifestation of this mycobacterial disease is determined by the host's immune system. Proper classification of the disease is of fundamental priority to determine accurate diagnosis followed by unerring treatments and management of the patients. Amidst several classifications of leprosy, the most widely accepted classification has been the one which was reported by Ridley and Jopling in the year 1966. As per their report, the classification was based mainly on immunological,

Stage	Description
Intermediate leprosy (IL)	It is the first stage of leprosy with few visible flat lesions
Tuberculoid leprosy (TT)	It is mainly characterised by fewer solitary skin lesions which are typically hypopigmented or erythematous macules
Borderline tuberculoid leprosy (BT)	Different grades of skin lesions with varied nerve involvements are found here
Borderline leprosy (BB)	Cutaneous lesions are characteristically reddish annular plaques with moderate numbness, swollen lymph glands having sharp interior and exterior borders
Borderline lepromatous leprosy (BL)	It is basically the skin condition characterised by numerous dimorphic flat lesions with raised bumps, nodules, and sometimes numb
Lepromatous leprosy (LL)	This type of leprosy is the most unfavourable clinical variant characterised by pale macules in the skin with no epithelioid cells in the lesions

Table 1.
Stages of leprosy [6].

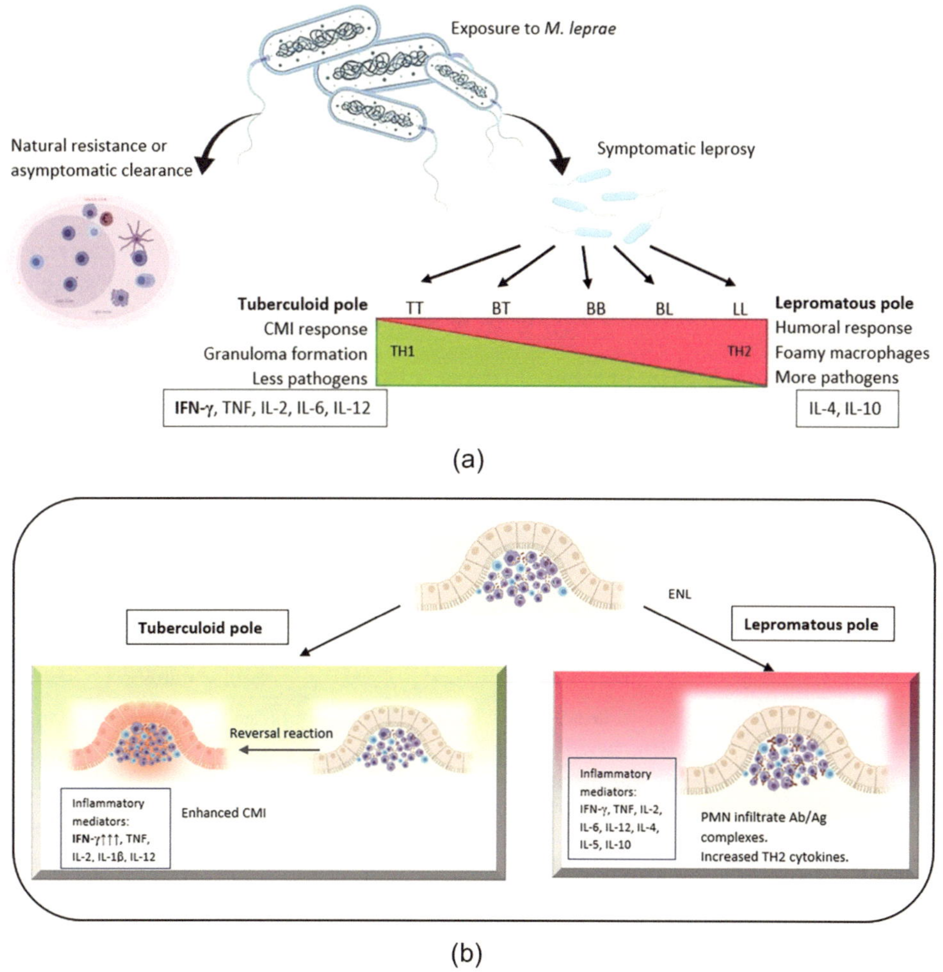

Figure 2.
(a) Effect on tuberculoid pole and lepromatous pole upon exposure to M. leprae *by the human immune system. (b) Response of human immune system at the tuberculoid pole and lepromatous pole under Erythema nodosum leprosum (ENL).*

whistopathological, and microbiological parameters and the immune status of the host [5]. There are six stages of leprosy with varying clinical symptoms (**Table 1**).

As many of the public health facilities might not have the technical setup to follow the above classifications, a comparatively simpler flowchart of classification is being followed. Using Ridley's bacterial index (BI) as a primary criterion, WHO in 1982 classified leprosy as multibacillary (infectious) and paucibacillary (non-infectious). Patients (BB, BL, LL) with BI $\geq$ 2 are classified as multibacillary. Besides, patients (TT and BT) with BI < 2 at all sites are classified under paucibacillary. Patients with TT and LL elicit different types of immune responses in the body (**Figure 2a**). Considering clinical and operational needs to avoid treatment inconveniences, smear-positive cases were grouped under multibacillary whereas smear-negative cases were considered paucibacillary [5]. There is an increase in the response of IFN-γ at the tuberculoid pole than at the lepromatous pole (**Figure 2b**) [7].

4. Epidemiology of leprosy

In tropical countries, leprosy is found to be endemic although there are still many cases in Southeast Asia, America, Africa, Eastern Pacific, and Western Mediterranean [8]. As many as 171,948 new cases of leprosy were recorded in early 2012 with a prevalence of 0.23 cases per 10,000 inhabitants [9].

From the 15 million people treated with multidrug (rifampicin, clofazimine, and dapsone) therapy against leprosy, around 2 million people have been prevented from developing disabilities [10]. Prevalence of leprosy fell from 620,638 cases in 2002 to 213,036 in 2009 [11]. In the year 2020, 127,558 new cases of leprosy were detected worldwide from 139 countries. Out of these, 8629 cases were found in children below 15 years.

India records the highest number of leprosy cases in the world. A study conducted in the state of Maharashtra in India showed three to nine cases of leprosy per 10,000 population and 30% of these newly detected cases were found in children [12].

Leprosy cases in Brazil are a major health problem. Brazil ranks second in the number of leprosy cases with a prevalence rate of 1.54 cases per 10,000 inhabitants. In 2011, there were 33,955 new cases out of which 61% were multibacillary (MB).

Year	Place	No. of cases
2002	Worldwide	6,20,638
2003	Brazil	4181 (detected in children under the age of 15 years)
2009	Worldwide	2,13,036
2011	Brazil	2420 (detected in children under the age of 15 years)
2011	Brazil	33,955 (cumulative)
2012	Southeast Asia, America, Africa, Eastern Pacific and Western Mediterranean	171,948
2022	Worldwide	1,27,558

Table 2.
Epidemiology of leprosy.

In 2003, 4181 cases were detected in children under the age of 15 years in Brazil which resulted in the detection coefficient of 7.98 per 100,000 inhabitants. The number of cases fell in 2011 with 2420 new cases resulting in a detection coefficient of 5.22 per 100,000 inhabitants (**Table 2**) [13].

WHO suggests that close disease surveillance for leprosy is necessary to eliminate the sources of infection and prevent the further spread of the disease. Tools are required for accurate, comparable grading practices. Although various instruments are available for measuring disabilities [14], their application in leprosy needs to be validated.

5. Histopathological features in leprosy

Histopathological analysis of LL skin, nasal swabs, and other tissues demonstrates that the majority of mycobacterial colonies are present inside macrophages. *M. leprae* was also noticed inside Schwann cells, pseudostratified epithelium, secretory glands, and ducts [15]. Analysis of skin lesions depicted that foamy macrophage is generally found in multibacillary (BB, BL, and LL) patients and epithelioid cells are usually present in paucibacillary (TT and BT) patients. During early and active infection, macrophages remain filled with granular eosinophilic cytoplasm colonised by a large number of bacilli. But in latent lesions, vacuolated cells are present with a foamy appearance [16]. In both paucibacillary and multibacillary leprosy, nerves are gradually destroyed and replaced by fibrous tissues [17].

Cell-mediated immunity (CMI) plays a major role in eradicating tuberculoid form or paucibacillary leprosy. CMI forms granulomas, destroying most of the mycobacteria, with traces of few remaining in the tissues. Skin and peripheral nerves face severe damage, but TT or the tuberculoid form progress slowly and the patient usually survives. Whereas there is a hike in humoral immunity for lepromatous form or multibacillary leprosy and the cell-mediated response is depressed, sometimes resulting in hypergammaglobulinemia. The mycobacteria are widely disseminated in

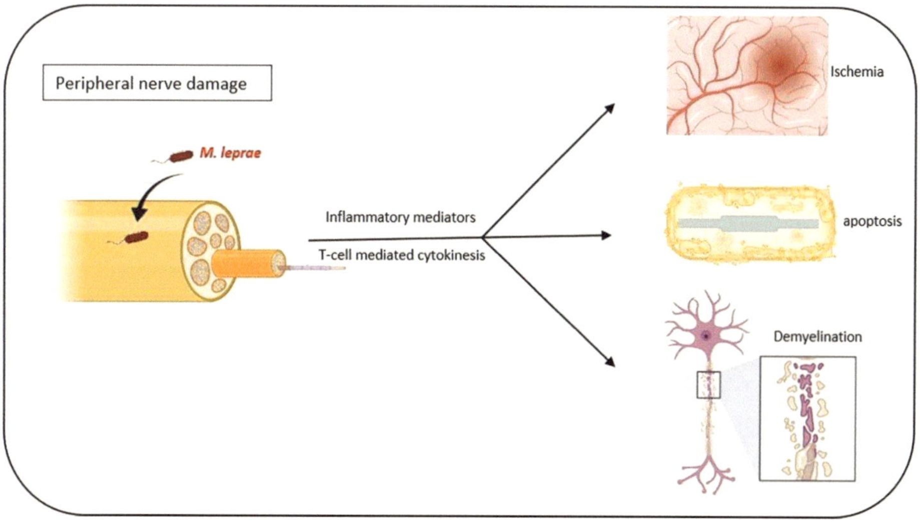

Figure 3.
Effects of peripheral nerve damage by M. leprae.

macrophages with the number reaching as high as 10^{10} per gram of tissue. LL causes disseminated infection of bones and cartilages with extensive nerve damage. Recent studies explained the fact that the macrophages present in lepromatous skin tissues are positive for adipose differentiation-related protein (ADRP). This further demonstrates that their foamy appearance is due to lipid bodies accumulation by *M. leprae* [18, 19].

Two types of leprosy reactions are noted in the patients. One is the reversal reaction. It is an acute inflammatory episode that occurs in skin and nerves in response to immediate activation of cellular immunity against the pathogen [20]. It affects dermis and Schwann cells in peripheral nerves causing demyelination, apoptosis, and ischaemia (**Figure 3**). Activated epithelioid macrophages are found in the reversal reaction lesions. Another is Erythema nodosum leprosum (ENL). It is clinically reported in approximately 50% of patients from lepromatous poles. It occurs primarily due to complex interaction between innate and CMI [21, 22]. ENL is marked by the infiltrate of neutrophils in the dermis and hypodermis, often accompanied by macrophages [23, 24].

6. Interferon-gamma (IFN-γ) in mycobacterial infection: a preamble

IFNs are a smaller subdivision of a larger class of proteins called cytokines, which are molecules used to establish communication between immune cells and non-immune cells to trigger the action of the immune system that helps in the eradication of pathogens [25, 26]. IFNs get their name because they have the capacity to "interfere" with viral replication and put a stop to further viral infections [26]. These not only help in the activation of immune cells but they also help to escalate host defences by increasing MHC (Major Histocompatibility Complex antigens) expression by causing an up-regulation in antigen expression markers. There exist more than 20 distinct IFN genes in animals and they are divided into three main classes: Type I, II, and III category. Out of these, IFN-γ is the sole member of the Type II IFN, mainly secreted by M1 effector macrophages, T-cells, and NK cells which are activated by interleukin-12. It is also referred to as the immune IFN [26].

IL-18 along with IL-1 is the key player in inducing IFN-γ production. IL-18 and IL-12 synergize with each other during the production of IFN-γ [27]. These IFNs block the proliferation of type-2 T helper cells; they are essentially released by type-1 T helper cells, cytotoxic T cells, macrophages, mucosal epithelial cells, and NK cells [28]. IFN-γ blocks the Th2 immune response system but furthers the Th1 immune response system (**Figure 4**) [29]. The released IFN-γ binds to the IFN-γ receptor protein complex (IFN-γ R) which is a heterodimer of two chains, IFNγR1 and IFNγR2 [26].

IFN-γ confers adaptive and innate immunity against bacterial, viral, and many protozoan diseases. It is a key stimulator of macrophages where it augments lysosome activity for effective management of bacterial burden. It helps to initiate binding and adhesion required for proper leukocyte migration. Not only does it help in priming alveolar macrophages against secondary bacterial infection, but it also increases the expression of class I MHC as well as class II MHC molecules through induction of antigen processing genes [30, 31].

IFN-γ interacts with the specific heterodimeric IFN-γ receptors (IFN-γR) located on target cells, such as macrophages, dendritic cells, and many other cell types. Similar to the other IFNs, IFN-γ also signals via the classical Janus kinase/signal transducers and activators of the transcription (JAK-STAT) signalling pathway (**Figure 5**).

Functional IFN-γR is made up of two ligand-binding α subunits, IFN-γR1 (drawn in black (**Figure 5**)), and two signal transducing β subunits, IFN-γR2

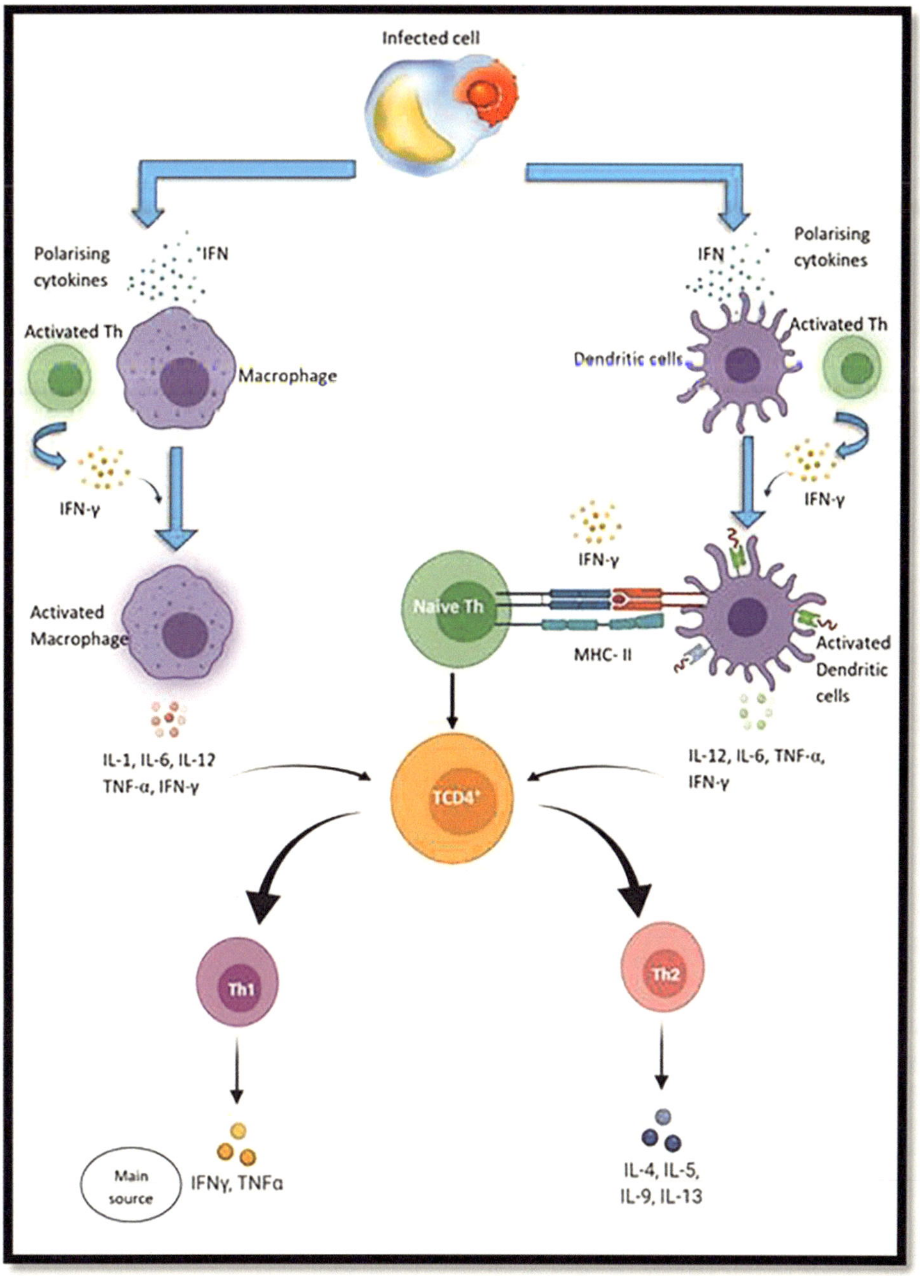

Figure 4.
Pathway of IFN-γ in the immune system in response to pathogen.

(drawn in red (**Figure 5**)). Both the receptor chains are classified under the class II cytokine receptor family. IFN-γ stimulates the heterodimerization of these two types of receptor chains, prior to its binding to IFN-γR1. However, such stimulation and binding happen only when the two mature IFN-γ monomers associate to form a biologically active homo-dimer [32]. The IFN-γR1 and IFN-γR2 subunits are associated with Janus Tyrosine Kinases, JAK1, and JAK2 [33, 34]. Followed

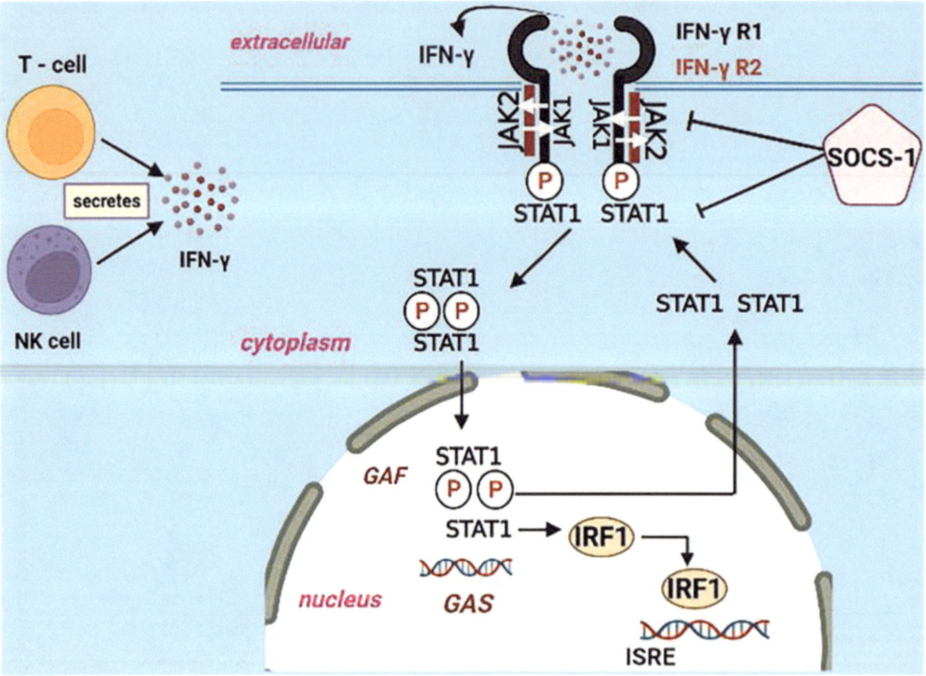

Figure 5.
Interferon-γ is mainly secreted by T-cell and NK cell signals via the classical Janus kinase/signal transducers and activators of the transcription (JAK/STAT) signalling pathway.

by IFN-γ binding, the two receptor subunits undergo cross-linking and auto-phosphorylation, and subsequent activation of JAK1 and JAK2 occurs [35]. The intracellular domains of IFN-γR1 contain binding motifs for JAK 1 and the Signal Transducer and Activator of Transcription protein called STAT-1, a latent cyto-plasmic transcription factor [28]. Phosphorylation of the STAT1 binding motif at tyrosine (Y) 440 residues promotes the recruitment of STAT1in the nucleus. Activated JAK2 is the major player in phosphorylation of mostly latent STAT-1 close to its C terminal region at Y701 [19, 36, 37]. Phosphorylated STAT-1 forms homo-dimers and subsequently detaches from the receptor and translocates into the nucleus to interact with the γ-activation site (GAS) elements at sequences like TTCN(2-4) GAA, within the promoter regions, to either stimulate or repress IFN-γ-regulated genes [38]. Therefore, IFN-γ-IFN-γR signalling induces or triggers several transcription factors like the IRF1, which plays key roles in regulating adaptive or innate immune responses, stimulating further transcription processes, and activating other transcription factors simultaneously.

Studies on IFN-γ or IFN-γR1 or IFN-γR2 deficient animals revealed that these animals are highly susceptible to a wide range of microbial and some viral pathogens [26]. Similar observations have been seen in humans presenting loss of function mutations in either the IFN-γR1 or IFN-γR2 chains. Such patients express poor immunity towards mildly virulent mycobacterium, early onset of its infection, and often death at a young age [28]. IFN-γ exercises significant roles in inflammatory responses and immune regulation. Infants having complaints of deficient IFN-γ production show inhibited neutrophil mobility and NK cell functioning [39].

7. Role of neutrophils

Neutrophils from LL patients with or without ENL release TNF and IL-8 when they are stimulated with *M. leprae* [40]. The apoptotic rate of ENL neutrophils is found to be greater when compared to LL patients and healthy people [40]. E-selectin and its ligands are the key molecules that mediate neutrophil recruitment to inflammatory sites. It is observed in the microarray analyses comparing skin lesions of LL patients and patients with ENL that an up-regulation of these cell movement genes happens in ENL [41]. Circulating neutrophils express CD64 on the cell surface during ENL. This phenomenon is absent in reversal reaction (RR). Non-reactional leprosy or healthy people have lower levels of CD64 expression on the cell surface [42]. Neutrophil function and adhesion to the endothelium increase with increased CD64 expression in cells *in vivo* [43–46].

LL is characterised by the presence of massive granulomas containing severely infected macrophages. This is due to the impairment of both RNI and ROS pathways in monocytes of LL patients. This deficiency is rectified by injecting recombinant human IFN-γ intradermally into lepromatous lesions, which resulted in measurable bacilli clearance from the lesions [47].

As a direct means of assessing the killing potential of infected macrophages against intracellular *M. leprae* is not available, *Toxoplasma gondii* death was utilised as an indicator of such microbicidal activity [48, 49]. Using the footpad model, Sibley and Krahenbuhl demonstrated that IFN-γ treatment does not activate *M. leprae*-burdened-footpad-granuloma-macrophages to inhibit *Toxoplasma gondii* [48]. This deficiency in activity is linked to an excess of intracellular Lipoarabinomannan (LAM) in macrophages [49], a significant ingredient of the mycobacterial cell wall and the primary carbohydrate-containing component identified by antibodies in TB and leprosy patients' sera [50].

8. Antimicrobial effects of IFN-γ against *M. tuberculosis* and *M. leprae*

The activation of cellular immunity and inflammation by IFN-γ is a characteristic of tuberculosis infection. Animals lacking either IFN-γ or IFNγ-R gene are vulnerable to mycobacterial infections, and this deficiency promotes fulminant mycobacterial growth and develops disseminated tuberculosis [51].

In IFN-γ treated macrophages, *M. tb* and *M. leprae* suppress MHC class II expression and antigen processing, dampening the pro-inflammatory and protective effects of IFN-γ. This effect is thought to be owing to inhibitory effects imposed on the chromatin remodelling of class II transactivator (CIITA) via the TLR2 and MAPK pathways, as well as restricted expression of CIITA in a TLR-dependent manner [52, 53]. *M. tb* inhibits downstream transcriptional responses caused by IFN-γ, though the proximal stages in IFNγ-R signalling, including STAT1 phosphorylation and dimerization, are unaffected. Diminished interaction of STAT1 dimers with related co-activators, such as cAMP response element-binding protein (CREB) and p300 in *M. tb* infected macrophages is ascribed to reduced IFN-γ induced responses [54]. Whereas *M. leprae* is found to negatively regulate macrophage-driven immune response by inducing high levels of MCP1 [55]. A high level of MCP1 expression is observed as a marker of nerve damage in leprosy [56]. On the other hand, TLR signalling has been connected to the generation of inhibitory responses to IFN-γ in mycobacterium-infected macrophages. TLR2 activation has been shown to suppress

IFN-γ responsive effects by stabilising and expressing a dominant-negative version of STAT1β. Higher level or stabilisation of STAT1β is found to be associated with a lowering of IFN-γ gene expression [53]. In cases of *M. avium* infection, higher amounts of STAT1β are produced, which lowers the gene expression of IFN-γ [53]. The pathogenicity by *M. leprae* triggers the host immune system to prepare for the defence against the same. The circulating monocytes identify and enter the site of the disease, where the local microenvironment activates the tissue macrophages to differentiate.

The disease lesions in TT are associated with well-organised granulomas with M1 macrophages. It expresses macrophage marker CD209 armed with antimicrobial effector function [57]. In LL patients the lesions are distinguished by disorganised granulomas containing macrophages. It is CD209/163++ but does not have antimicrobial activity. These macrophages accumulate host-derived lipids which favours the *M. leprae* growth and are identified as M2 macrophages [58].

The study of the pathogenicity of leprosy reported predominancy of M1 and M2 type macrophages in the self-limited form of lesions and the progressive form of lesions, respectively. In normal situations unstimulated endothelial cells (EC) trigger monocytes to differentiate into M2 macrophages which are phagocytic in nature [59, 60]. Further biochemical screening analysis depicted that when IFN-γ acts upon EC, it differentiates monocytes into M1 macrophages. It has been hypothesised that in a stable micro-environment if the infection is below a detectable level, then

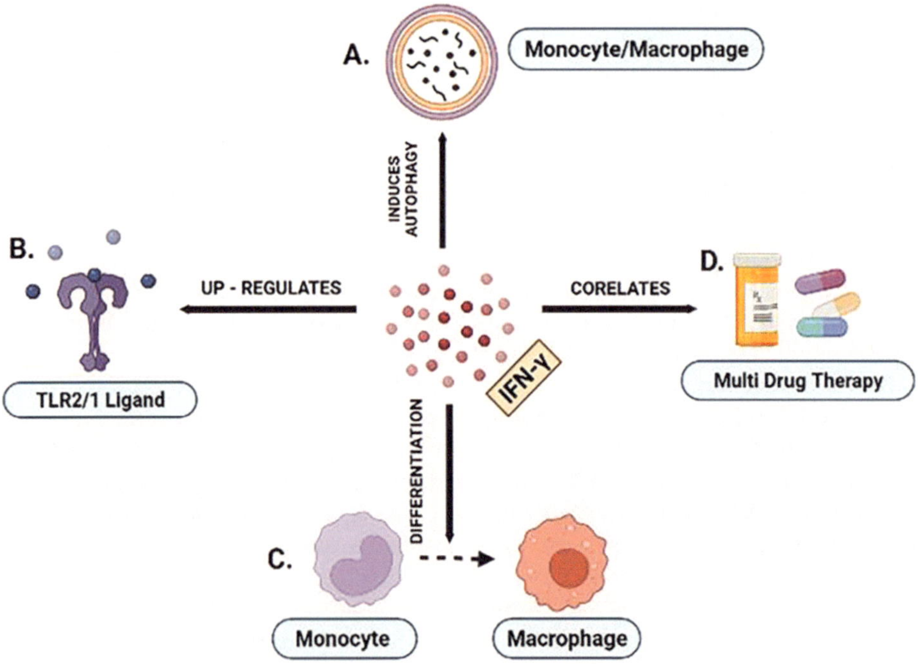

Figure 6.
Interferon-gamma in leprosy. (A) Human monocytes and macrophages when treated with IFN-γ induce autophagy of M. leprae *with the assistance of vitamin D. (B) IFN-γ up-regulates TLR2/1 Ligand inducing antimicrobial peptide expression by activating CYP27BI. (C) In the pathogenic micro-environment of* M. leprae, *IFN-γ provokes EC to facilitate monocyte differentiation to CD209 + CD163 M1 macrophages which is associated with host defence for immediate antimicrobial activity. (D) IFN-γ induces cathelicidin in* M. leprae*-infected monocytes in the presence of vitamin D enriched serum which is strongly correlated during multidrug therapy of leprosy.*

EC signals monocytes to differentiate into M2 macrophages. But in the pathogenic micro-environment, IFN-γ provokes EC to instruct monocyte differentiation into M1 macrophages for immediate antimicrobial activity (**Figure 6**) [61]. IFN-γ primed T cells induce antimicrobial peptide gene expression in monocytes and macrophages in response to mycobacteria [62].

Many research groups studied various immune response combinations of cytokines, EC, and macrophages in leprosy. Most of the cytokines used in the pre treatment, cumulatively triggered EC to signal monocyte differentiation to M2 macrophages. Only IFN-γ-treated EC facilitated monocyte differentiation to CD209 +/ CD163 M1 macrophages which is associated with host defence [56]. But monocyte differentiation to M1 macrophages ceases when IFN-γ acts directly upon the monocyte in absence of EC. The M1 macrophages induced by IFN-γ-treated EC were clinically more active than M2 macrophages induced by resting EC. The former was reported to (i) accumulate less oxidised LDL, to prevent the nourishment of *M. leprae*. (ii) Initiates mass production of pro-inflammatory cytokines in response to mycobacterial TLR2/1 ligand [56] (iii) the level of vitamin D antimicrobial pathway genes, for instance: Cyp27b1, VDR, and cathelicidin are all elevated greatly against the pathogen [1, 53, 63, 64]. This emphasises that IFN-γ contributes to an active defence mechanism against *M. leprae* and is a potent inflammatory mediator which stimulates an extensive gene program in EC as it fails to directly trigger monocytes differentiation to respective macrophages expressing CD209 phenotype and antimicrobial function [65, 66].

Thereafter, IFN-γ induces CYP27B1-hydroxylase in monocytes and macrophages which converts 25-hydroxyvitamin D (25D) to bioactive 1,25-dihydroxy vitamin D (1,25D) [66]. IFN-γ up-regulates TLR2/1 ligand inducing antimicrobial peptide expression by activating CYP27BI (**Figure 6**) [67]. Antagonistically IFN-γ down-regulates the CYP24 gene which instigates the production of antimicrobial peptides. *M. leprae* evades the macrophage antimicrobial response successfully by obstructing phagosome maturation and phagolysosomal fusion [68–70]. An efficient host defence mechanism to overcome this barrier is autophagy which creates autophagosomes and their corresponding fusion with lysosomes [71–74]. IFN-γ plays an active role in this autophagy (**Figure 6**). IFN alone or in a combination with vitamin D induces autophagy in human monocytes and macrophages [75]. IFN-γ induces the secretion of cathelicidin in *M. leprae*-infected monocytes in the presence of vitamin D enriched serum. Vitamin D and cathelicidin levels can both be strongly correlated during multidrug therapy of leprosy. This combination strengthens the immune system of the host against leprosy suggesting the concerted antileprosy function of vitamin D and IFN-γ (**Figure 6**) [76].

In another study, it is reported that IFN-γ-mediated autophagy in macrophages leads to control in *M. leprae* counts in TT leprosy. Whereas in LL leprosy, BCL2 mediated block in autophagy results in augmented IFN-γ, reversal reactions, and tissue damage [77]. Therefore, the use of IFN-γ as a part of leprosy therapy could be a challenging endeavour as it could do more harm than curing the disease. To date, only one noteworthy effort has been made to this effect where intradermal injection of IFN-γ reduced BI of leprosy in some cases but patients are four-fold more prone to develop ENL/Type II reactions, offsetting the usefulness of IFN-γ therapy [78].

9. Conclusion

Leprosy still affects a large number of people worldwide causing several forms of disabilities in them. IFN-γ, a type II IFN, possesses antimicrobial and antiviral activities. This activates endothelial cells to facilitate the differentiation of monocytes into M1 macrophages which are phagocytic in nature in a pathogenic microenvironment. IFN-γ activates T cells to induce antimicrobial gene expression in response to *M. leprae*. IFN primed endothelial cells trigger the differentiation of macrophages which prevented the growth of *M. leprae*. It is also seen that IFN-γ played an important role in escalating antimicrobial and anti-inflammatory pathways in conjunction with vitamin D to eliminate *M. leprae.* IFN-γ also plays an active role in autophagy to eliminate the pathogen suggesting that IFN-γ contributes to the host defence against leprosy. But, as a therapeutic agent, it has not been successfully used so far. A better understanding of leprosy immunology might help us to overcome this limitation for the application of IFN-γ in leprosy therapy.

Acknowledgements

PG and AB thank Amity University Kolkata for organisational support.

Funding

None.

Author details

Pragya Santra, Prama Ghosh, Soumyadeep Ghosh, Anwesha Behera, Oyendrilla Mitra, Ishanee Das Sharma, Diya Adhikary, Asesh Banerjee* and Prabuddha Gupta*
Amity Institute of Biotechnology, Amity University, Kolkata, West Bengal, India

*Address all correspondence to: abanerjee1@kol.amity.edu and pgupta@kol.amity.edu

References

[1] Pinheiro RO et al. Innate immune responses in leprosy. Frontiers in Immunology. 2018;**9**:1-15. DOI: 10.3389/fimmu.2018.00518

[2] Ploemacher T, Faber WR, Menke H, Rutten V, Pieters T. Reservoirs and transmission routes of leprosy: A systematic review. PLoS Neglected Tropical Diseases. 2020;**14**(4):e0008276. DOI: 10.1371/journal.pntd.0008276

[3] Silva CAM et al. Interaction of *Mycobacterium leprae* with human airway epithelial cells: Adherence, entry, survival, and identification of potential adhesions by surface proteome analysis. Infection Immunology. 2013;**81**(7):2645-2659. DOI: 10.1128/IAI.00147-13

[4] De Lima CS et al. Heparin-binding hemagglutinin (HBHA) of *Mycobacterium leprae* is expressed during infection and enhances bacterial adherence to epithelial cells. FEMS Microbiology Letters. 2009;**292**(2):162-169. DOI: 10.1111/j.1574-6968.2009.01488.x

[5] Parkash O. Classification of leprosy into multibacillary and paucibacillary groups: An analysis. FEMS Immunology and Medical Microbiology. 2009;**55**(1):1-5. DOI: 10.1111/j.1574-695X.2008.00491.x

[6] Modlin RL. Th1-Th2 paradigm: Insights from leprosy. The Journal of Investigative Dermatology. 1994;**102**(6):828-832

[7] Ridley MJ, Russell D. An immunoperoxidase study of immunological factors in high immune and low resistance granulomas in leprosy. The Journal of Pathology. 1982;**137**(2):149-157

[8] Lastória JC, de Abreu MAMM. Leprosy: Review of the epidemiological, clinical, and etiopathogenic aspects—Part 1. Anais Brasileiros de Dermatologia. 2014;**89**(2):205-218. DOI: 10.1590/abd1806-4841.20142450

[9] Rao P, Suneetha S. Current situation of leprosy in India and its future implications. Indian Dermatology Online Journal. 20189(2):83-89. DOI: 10.4103/idoj.idoj_282_17

[10] Scollard DM, Adams LB, Gillis TP, Krahenbuhl JL, Truman RW, Williams DL. The continuing challenges of leprosy. Clinical Microbiology Reviews. 2006;**19**(2):338-381. DOI: 10.1128/CMR.19.2

[11] Region EA, Region M. Global leprosy situation, 2012 = Situation mondiale de la lèpre, 2012. Weekly Epidemiological Records. 2012;**87**(34):317-328

[12] Rodrigues LC, Lockwood DNJ. Leprosy now: Epidemiology, progress, challenges, and research gaps. The Lancet Infectious Diseases. 2011;**11**(6):464-470. DOI: 10.1016/S1473-3099(11)70006-8

[13] Santos SD, Penna GO, Costa N, Natividade MS, Teixeira MG. Leprosy in children and adolescents under 15 years old in an urban centre in Brazil. Memorias do Instituto Oswaldo Cruz. 2016;**111**(6):359-364

[14] Van Brakel WH, Officer A. Approaches and tools for measuring disability in low and middle-income countries. Leprosy Review. 2008;**79**(1):50-64. DOI: 10.47276/lr.79.1.50

[15] Mcdougall AC, Rees RJW, Weddell AGM, Wmdi Kanan M. The Histopathology of Lepromatous Leprosyin the Nose. London: National Institute for Medical Research; 1975 **115**(4):215-226

[16] Desikan KV, Iyer CG. The distribution of *Mycobacterium leprae* in different structures of the skin. Leprosy Review. 1972;**43**(1):30-37

[17] Petito RB et al. Transforming growth factor-β1 may be a key mediator of the fibrogenic properties of neural cells in leprosy. Journal of Neuropathology and Experimental Neurology. 2013;**72**(4):351-365

[18] Mattos KA et al. Lipid droplet formation in leprosy: Toll-like receptor-regulated organelles involved in eicosanoid formation and *Mycobacterium leprae* pathogenesis. Journal of Leukocyte Biology. 2010;**87**(3):371-384. DOI: 10.1189/jlb.0609433

[19] de Mattos KA, Sarno EN, Pessolani MCV, Bozza PT. Deciphering the contribution of lipid droplets in leprosy: Multifunctional organelles with roles in *Mycobacterium leprae* pathogenesis. Memórias do Instituto Oswaldo Cruz. 2012;**107**(SUPPL):1. DOI: 10.1590/S0074-02762012000900023

[20] Kamath S, Vaccaro SA, Rea TH, Ochoa MT. Recognizing and managing the immunologic reactions in leprosy. Journal of the American Academy of Dermatology. 2014;**71**(4):795-803. DOI: 10.1016/j.jaad.2014.03.034

[21] Andrade PR, Amadeu TP, Nery JA, Pinheiro RO, Sarno EN. CD123, the plasmacytoid dendritic cell phenotypic marker, is abundant in leprosy type 1 reaction. British Journal of Dermatology. 2015;**172**(1):268-271. DOI: 10.1111/bjd.13430

[22] de Souza Sales J et al. The role of indoleamine 2, 3-dioxygenase in lepromatous leprosy immunosuppression. Clinical Experimental in Immunology. 2011;**165**(2): 251-263

[23] Cuevas J, Rodríguez-Peralto JL, Carrillo R, Contreras F. Erythema nodosum leprosum: Reactional leprosy. Seminars in Cutaneous Medicine and Surgery. 2007;**26**(2):126-130. DOI: 10.1016/j.sder.2007.02.010

[24] Schmitz V et al. Expression of CD64 on circulating neutrophils favoring systemic inflammatory status in erythema nodosum leprosum. PLoS Neglected Tropical Diseases. 2016;**10**(8):1-18

[25] De Andrea M, Ravera R, Gioia D, Gariglio M, Landolfo S. The interferon system: An overview. European Journal of Paediatric Neurology. 2002;**6**: A41-A46

[26] Parkin J, Cohen B. "An overview of the immune system". Lancet. Jun 2001;357:1777-1789

[27] Shtrichman R, Samuel CE. The role of gamma interferon in antimicrobial immunity. Current Opinion in Microbiology;**4**:251-259

[28] Schroder K, Hertzog PJ, Ravasi T, Hume DA. Interferon-γ: An overview of signals, mechanisms and functions. Journal of Leukocyte Biology. 2004;**75**(2):163-189. DOI: 10.1189/jlb.0603252

[29] Kidd P. Th1/Th2 Balance: The Hypothesis, its Limitations, and Implications for Health and Disease. 2003;**8**(3):223-246

[30] Hoyer FF et al. Tissue-specific macrophage responses to remote injury impact the outcome of subsequent local immune challenge. Immunity. 2019;**51**(5):899-914.e7. DOI: 10.1016/j.immuni.2019.10.010

[31] Yao Y et al. Induction of autonomous memory alveolar macrophages requires

T cell help and is critical to trained immunity. Cell. 2018;**175**(6):1634-1650.e17. DOI: 10.1016/j.cell.2018.09.042

[32] Krause CD et al. Seeing the light: Preassembly and ligand-induced changes of the interferon gamma receptor complex in cells. Molecular & Cellular Proteomics. 2002;**1**(10):805-815. DOI: 10.1074/mcp.M200065-MCP200

[33] Chen J, Baig E, Fish EN. Diversity and Relatedness Among the Type I Interferons. 2004;**24**(12):687-698

[34] Bach EA, Aguet M, Schreiber RD. THE IFNγ RECEPTOR: A Paradigm for Cytokine Receptor Signaling. 1997;**15**:563-591

[35] Platanias LC. Mechanisms of type-I- and type-II-interferon-mediated signalling. Nature Reviews Immunology. 2005;**5**(5):375-386. DOI: 10.1038/nri1604

[36] Schindler C. Cytokines and JAK-STAT Signaling. 1999:**253**(1):7-14

[37] Briscoe J et al. Kinase-negative mutants of JAK1 can sustain interferon-γ-inducible gene expression but not an antiviral state. The EMBO Journal. 1996;**15**(4):799-809. DOI: 10.1002/j.1460-2075.1996.tb00415.x

[38] Ann Liebert M, Decker T, Kovarik P, Meinke A, Biocenter V. GAS Elements: A Few Nucleotides with a Major Impact on Cytokine-Induced Gene Expression. 1997:121-134

[39] Jolles S, Kaveri SV, Orange J. Clinical and experimental immunology: Foreword. Clinical and Experimental Immunology. 2009;**158**:1

[40] Oliveira RB, Moraes MO, Oliveira EB, Sarno EN, Nery JA, Sampaio EP. Neutrophils isolated from leprosy patients release TNF-alpha and exhibit accelerated apoptosis in vitro. Journal of Leukocyte Biology. 1999;**65**:364-371. DOI: 10.1002/jlb.65.3.364

[41] Lee DJ, Li H, Ochoa MT, Tanaka M, Carbone RJ, Damoiseaux R, et al. Integrated pathways for neutrophil recruitment and inflammation in leprosy. The Journal of Infectious Diseases. 2010;**201**:558-569. DOI: 10.1086/650318

[42] Schmitz V, Prata RB, Barbosa MG, Mendes MA, Brandão SS, Amadeu TP, et al. Expression of CD64 on circulating neutrophils favoring systemic inflammatory status in erythema nodosum leprosum. PLoS Neglected Tropical Diseases. 2016;**10**:e0004955. DOI: 10.1371/journal.pntd.0004955

[43] Höglund M, Håkansson L, Venge P. Effects of in vivo administration of G-CSF on neutrophil functions in healthy volunteers. European Journal of Haematology. 1997;**58**:195-202. DOI: 10.1111/j.1600-0609.1997.tb00947.x

[44] Turzanski J, Crouch SP, Fletcher J, Hunter A. Ex vivo neutrophil function in response to three different doses of glycosylated rHuG-CSF (lenograstim). British Journal of Haematology. 1997;**96**:46-54. DOI: 10.1046/j.1365-2141.1997.d01-2000.x

[45] Repp R, Valerius T, Sendler A, Gramatzki M, Iro H, Kalden JR, et al. Neutrophils express the high affinity receptor for IgG (Fc gamma RI, CD64) after in vivo application of recombinant human granulocyte colony-stimulating factor. Blood. 1991;**78**:885-889

[46] Fadlon E, Vordermeier S, Pearson TC, Mire-Sluis AR, Dumonde DC, Phillips J, et al. Blood polymorphonuclear leukocytes from the majority of sickle cell patients in the crisis phase of the disease show enhanced adhesion to vascular

endothelium and increased expression of CD64. Blood. 1998;**91**:266-274

[47] Denis M. Tumor necrosis factor and granulocyte macrophage-colony stimulating factor stimulate human macrophages to restrict growth of virulent *Mycobacterium avium* and to kill avirulent *M. avium*: Killing effector mechanism depends on the generation of reactive nitro. Journal of Leukocyte Biology. 1991;**49**(4):380-387. DOI: 10.1002/jlb.49.4.380

[48] Sibley LD, Krahenbuhl JL. *Mycobacterium leprae*-burdened macrophages are refractory to activation by gamma interferon. Infectious Immunology. 1987;**55**(2):446-450

[49] Sibley LD, Krahenbuhl JL. Defective activation of granuloma macrophages from *Mycobacterium leprae*-infected nude mice. Journal of Leukocyte Biology. 1988;**43**(1):60-66. DOI: 10.1002/jlb.43.1.60

[50] Czarniecki CW, Sonnenfeld G. Interferon-gamma and resistance to bacterial infections. APMIS. 1993; **101**(1-6):1-17. DOI: 10.1111/j.1699-0463.1993.tb00073.x

[51] Bozzano F, Marras F, De Maria A. Immunology of tuberculosis. Mediterranean Journal of Hematology and Infectious Diseases. 2014;**6**(1) e2014027. DOI: 10.4084/MJHID.2014.027

[52] Harding CV, Boom WH. Regulation of antigen presentation by *Mycobacterium tuberculosis*: A role for Toll-like receptors. Nature Reviews. Microbiology. 2010;**8**(4):296-307. DOI: 10.1038/nrmicro2321

[53] Pennini ME, Pai RK, Schultz DC, Boom WH, Harding CV. *Mycobacterium tuberculosis* 19-kDa lipoprotein inhibits IFN-γ-Induced chromatin remodeling of MHC2TA by TLR2 and MAPK signaling. Journal of Immunology. 2006;**176**(7):4323-4330. DOI: 10.4049/jimmunol.176.7.4323

[54] Kak G, Raza M, Tiwari BK. Interferon-gamma (IFN-γ): Exploring its implications in infectious diseases. Biomolecular Concepts. 2018;**9**(1):64-79

[55] Sinsimer D, Fallows D, Peixoto B, Krahenbuhl J, Kaplan G, Manca C. *Mycobacterium leprae* actively modulates the cytokine response in naïve human monocytes. Infectious in Immunology. 2010;**78**(1):293-300

[56] Medeiros MF et al. CXCL10, MCP-1, and other immunologic markers involved in neural leprosy. Applied Immunohistochemistry and Molecular Morphology. 2015;**23**(3):220-229

[57] Montoya D et al. Divergence of macrophage phagocytic and antimicrobial programs in leprosy. Cell Host & Microbe. 2009;**6**(4):343-353. DOI: 10.1016/J.CHOM.2009.09.002

[58] Cruz D et al. Host-derived oxidized phospholipids and HDL regulate innate immunity in human leprosy. The Journal of Clinical Investigation. 2008;**118**(8):2917-2928. DOI: 10.1172/JCI34189

[59] Randolph GJ, Beaulieu S, Lebecque S, Steinman RM, Muller WA. Differentiation of monocytes into dendritic cells in a model of transendothelial trafficking. Science. 1998;**282**(5388)480-483. DOI: 10.1126/science.282.5388.480

[60] He H et al. Endothelial cells provide an instructive niche for the differentiation and functional polarization of M2-like macrophages. Blood. 2012;**120**(15):3152-3162. DOI: 10.1182/blood-2012-04-422758

[61] Kibbie J et al. Jagged1 instructs macrophage differentiation in leprosy. PLoS Pathogens. 2016;**12**(8)e1005808

[62] Fabri M et al. Vitamin-D is required for IFN-γ-mediated antimicrobial activity of human macrophages. Science Translational Medicine. 2011;**3**(104):102. DOI: 10.1126/scitranslmed.3003045

[63] Krausgruber T et al. IRF5 promotes inflammatory macrophage polarization and T H1-TH17 responses. Nature Immunology. 2011;**12**(3):231-238. DOI: 10.1038/ni.1990

[64] Saha B, Jyothi Prasanna S, Chandrasekar B, Nandi D. Gene modulation and immunoregulatory roles of Interferonγ. Cytokine. 2010;**50**(1)1-14. DOI: 10.1016/j.cyto.2009.11.021

[65] Indraccolo S et al. Identification of genes selectively regulated by IFNs in endothelial cells. Journal of Immunology. 2007;**178**(2):1122-1135. DOI: 10.4049/jimmunol.178.2.1122

[66] Adams JS, Gacad MA. Characterization of 1α-hydroxylation of vitamin-D3 sterols by cultured alveolar macrophages from patients with sarcoidosis. Journal of Experimental Medicine. 1985;**161**(4)755-765

[67] Edfeldt K et al. T-cell cytokines differentially control human monocyte antimicrobial responses by regulating vitamin-D metabolism. Proceedings of the National Academy of Science USA. 2010;**107**(52)22593-22598

[68] Sturgill-Koszycki S et al. Lack of acidification in Mycobacterium phagosomes produced by exclusion of the vesicular proton-ATPase. Science (80-.). 1994;**263**(5147)678-681

[69] Armstrong JA, Hart D. Response of cultured macrophages to *Mycobacterium Tuberculosis*, with observations on fusion of lysosomes with phagosomes. Journal of Experimental Medicine. 1971;**134**(3)713-740

[70] McDonough KA, Kress Y, Bloom BR. Pathogenesis of tuberculosis: Interaction of *Mycobacterium tuberculosis* with macrophages. Infections in Immunology. 1993;**61**(7)2763-2773

[71] Gutierrez MG, Master SS, Singh SB, Taylor GA, Colombo MI, Deretic V. Autophagy is a defense mechanism inhibiting BCG and *Mycobacterium tuberculosis* survival in infected macrophages. Cell. 2004;**119**(6)753-766. DOI: 10.1016/j.cell.2004.11.038

[72] Singh SB, Davis AS, Taylor GA, Deretic V. Human IRGM induces autophagy to eliminate intracellular mycobacteria. Science (80-.). 2006;**313**(5792)1438-1441

[73] Alonso S, Pethe K, Russell DG, Purdy GE. Lysosomal killing of Mycobacterium mediated by ubiquitin-derived peptides is enhanced by autophagy. Proceedings of the National Academy of Science USA. 2007;**104**(14)6031-6036. DOI: 10.1073/pnas.0700036104

[74] Tiwari S, Choi HP, Matsuzawa T, Pypaert M, MacMicking JD. Targeting of the GTPase Irgm1 to the phagosomal membrane via PtdIns(3,4)P2 and PtdIns(3,4,5)P3 promotes immunity to mycobacteria. Nature Immunology. 2009;**10**(8)907-917

[75] Kabeya Y et al. LC3, a mammalian homologue of yeast Apg8p, is localized in autophagosome membranes after processing. EMBO Journal. 2000;**19**(21)5720-5728

[76] Grossi de Oliveira AL et al. Hypovitaminosis D and reduced

cathelicidin are strongly correlated during the multidrug therapy against leprosy. Microbial Pathogenesis. 2020;**147**:104373. DOI: 10.1016/j.micpath.2020.104373

[77] de Silva A et al. Autophagy is an innate mechanism associated with leprosy polarization. PLoS Pathogens. 2017;**13**(1)e1006103

[78] Sampaio EP, Moreira AL, Sarno EN, Malta AM, Kaplan G. Prolonged treatment with recombinant interferon 3, induces erythema nodosum leprosum in lepromatous leprosy patients. Journal of Experimental Medical. 1992;**175**(6)1729-1737

Section 4

Interferon in Clinical Practice

Chapter 5

Perspective Chapter: The Role of Interferon Gamma in Clinical Medicine

Irina A. Rakityanskaya, Tat'jana S. Ryabova, Anastasija A. Kalashnikova, Goar S. Balasaniants, Andrej D. Kaprin, Feliks I. Ershov, Vera V. Kir'janova, Tat'jana B. Korzhenevskaja, Denis V. Barbinov, Andrej V. Ignatovskij, Ljudmila Y. Grivtsova, Valentina G. Isaeva, Natal'ja A. Falaleeva, Alisa I. Gil', Svetlana A. Berns, Natal'ja V. Vasil'eva, Julija V. Dolgo-Saburova, Elena V. Shagdileeva, Ekaterina V. Frolova and Nadezhda S. Astanina

Abstract

Interferon gamma (IFN-γ) is one of the key factors of both innate and adaptive immune response that promotes differentiation of naive CD4+ cells into effector Th1 T cells producing the main mediators of cellular immunity against viral and intracellular bacterial infections, and specific cytotoxic immunity through the interaction of T cells with antigen-presenting cells and macrophage activation. The clinical importance of IFN-γ includes its medical use to treat and prevent various viral and bacterial infections. IFN-γ has a direct antiviral effect on infected cells, activates local infiltrating dendritic cells, macrophages and NK cells, modulates the differentiation and maturation of T and B cells, and enhances inflammation and antiviral functions. Immunoregulatory effect of IFN-γ plays one of the essential roles in the regulation of adaptive immune response in patients with tuberculosis infection and cancer. Producing IFN-γ by T cells increases the efficiency of infiltrated phagocytic cells, by stimulating NO and maintaining local host defense during tuberculosis infection. The direct antitumor effect of IFN-γ revealed in several experimental models has numerous mechanisms for the effect of development. IFN-γ has crucial potential for enhancing any antiviral, antimycobacterial, and specific antitumor therapies.

Keywords: interferon gamma, macrophage, infectious pathogen, cytokine receptor, adaptive immunity, innate immunity, dendritic cell, tuberculosis, tumor

1. Introduction

Interferon gamma (IFN-γ) is the only member of the type II interferon family, which was discovered and described by E. Frederick Wheelock in 1965 as produced in vitro by leukocytes after their stimulation with phytohemagglutinin (Phaseolus vulgaris extract) and inducing antiviral activity. The physicochemical and biological properties of this virus inhibitor are similar to those of interferon induced by the Newcastle disease virus, except for instability at pH 2 and 10 and at 56°C [1]. IFN-γ is a protein encoded by the IFNG gene, consisting of two polypeptide chains linked in an antiparallel manner [2]. In peripheral blood, IFN-γ is present in three fractions with different molecular weights. One fraction is the active free form of IFN-γ and the other two are mature IFN-γ molecules. A fully synthesized protein is glycosylated at amino acid sites where the level of glycosylation determines the final weight of certain fractions [3, 4]. Glycosylation prevents the degradation of IFN-γ by proteinases increasing its half-life and prolonging the effects mediated by IFN-γ [5]. "Immune" interferon, also called IFN-γ, is a highly pleiotropic cytokine secreted not in response to infection, but indirectly by mitogen-activated T cells and natural killer (NK) cells, the primary producers of IFN-γ during both innate, and adaptive phases of the immune response.

2. General production and signaling pathways

2.1 Adaptive or innate immunity

IFN-γ is produced by NK and natural killer T cells (NKT) of innate immunity, gamma-delta T cells, and B cells. CD8+ and CD4+ T cells are the main paracrine sources of IFN-γ during the adaptive immune response [6]. Professional antigen presenting cells (APCs) have also been proven to secrete IFN-γ. The production of IFN-γ by monocytes/macrophages, dendritic cells acting locally, is important in cell activation [7, 8]. Normally, in the early stages of the host immune response, IFN-γ production by NK cells, CD4 + T (Th1) cells, and CD8 + T cells is aimed at improving antigen recognition in APCs, such as macrophages and dendritic cells.

IFN-γ is one of the key cytokines that promote differentiation of naive CD4+ cells into effector Th1 T cells that produce the main mediators of cellular immunity against viral and intracellular bacterial infections [9]. IFN-γ is the main product of Th1 cells and drives Th1 effector mechanisms: a) innate cell-mediated immunity (through activation of NK cell effector functions); b) specific cytotoxic immunity (through the interaction of T cells with APCs); c) activation of macrophages.

IFN-γ increases the content of lymphocytes and leads to their long-term persistence in the tissue, induces the activation of the cascade of complement components and an acute phase response, plays a role in switching the production of the IgG class, and has a direct antiviral effect [7]. When activated, almost all CD8+ T cells, NK cells, and Th1 lymphocytes produce IFN-γ, stimulating cytokine activity and increasing the expansion of low avid NK cells. Among all the interferons/cytokines of the Th1 response, IFN-γ correlates most strongly with the Th1 response.

CD4+ Th1 are the main source of IFN-γ, determined by the secretion of IL-12, IL-2 and IFN-γ, as well as the expression of T-bet, which is a transcription factor of the T-box family, encoded by the TBX21 gene, and plays the role of a promoter of IFN-γ synthesis [10, 11]. The expression level of T-bet correlates with the production of IFN-γ in Th1 and NK cells. Thus, IFN-γ is produced in response to multiple stimulants from the tissue-specific environment.

Classical inducers of IFN-γ production are IL-12 and IL-18. They activate IFN-γ production by NK cells and T cells [12, 13]. IL-12 has a powerful immunomodulatory effect on innate and adaptive immune cells. IL-12 is secreted as a biologically active 70 kDa heterodimer, consisting of disulfide-linked alpha (p35) and beta (p40) subunits [14]. Gene expression of IL-12 and IFN-γ is coordinated (i.e., IL-12 induces IFN-γ and IFN-γ induces IL-12). Binding of IL-12 to its relative heterodimeric receptor, IL12RB1/2, induces signaling through Jak-mediated phosphorylation of STAT4. Signaling through STAT4 induces IFN-γ expression [15]. Many of the effects mediated by IL-12 are due to the inducible expression of IFN-γ and the shift of CD4+ T cells towards the Th1 phenotype. Synergism between IL-12 and IL-18 has been shown to significantly induce IFN-γ in B cells [16]. IL-12 and other cytokines enhance T cell CTL activity by increasing sensitivity to weak or self-antigens. Together, IFN-γ and IL-12 generate a very strong Th1 response. Th1 cell-mediated cellular immunity and Th2 cell-mediated humoral immunity are modulated by IFN-γ, which influences the differentiation of naive T cells into Th1 or Th2 cells. Induction of IFN-γ in T cells initiates a positive feedback loop, as a result APCs sensitive to IFN-γ are primed to produce additional amounts of IL-12 [17]. IFN-γ blocks the production of IL-4, an inducer of Th2 cell differentiation and proliferation. The synergistic effect of IL-21, IL-18 and IL-15 enhances the production of IFN-γ. IL-15 is the strongest regulator of IFN-γ production compared to IL-21 in human NK and T cells. The cytokines IL-15 and IL-18 are produced by macrophages, while IL-21 is mainly produced by activated T cells. IL-24 or MDA-7 (Melanoma differentiation associated 7) can also activate the production of IFN-γ secreted by activated T-lymphocytes and monocytes and belongs to the IL-10 family [18]. IFN-γ increases the expression of HLA (major histocompatibility complex) class I and II antigen by increasing the expression of subunits increasing the expression and activity of proteasomes, and resulting in increased host sensitivity to an infectious pathogen and ability to identify and respond to this pathogen [19].

2.2 Interferon gamma crossroads

IFN-γ triggers antiviral and adaptive immune responses through the Janus kinase (Jak) and signal transducer and transcriptional activator (STAT) (Jak–STAT) signaling pathway which are the most studied intracellular signaling pathway. After IFN-γ binding and receptor dimerization, Jak1 and Jak2 are activated, which increases their catalytic activity and phosphorylation of the main target, STAT1. Phosphorylated STAT molecules get dimerized and transported to the nucleus, where they bind to the corresponding regulatory gene sequences and trigger their transcription [20]. In this *canonical signaling pathway*, IFN-γ get dimerized and binds to both IFN-γ receptors, which consist of two different ligand-binding chains: a high-affinity and highly expressed IFN-γ-R1 (α), and two signal-transforming low-affinity IFN-γR2 (β) with corresponding signaling mechanisms. The IFN-γR1 and IFN-γR2 chains belong to the family of class II cytokine receptors. The ligand-binding IFN-γ subunit IFN-γR1 and the assisting subunit IFN-γR2 correspond to chromosomes 6 and 21 in humans [21]. These subunits are intracellularly associated with the kinases of the Jak family, Jak1 and Jak2, respectively. Jak-1 interacts with the IFN-γR1 receptor subunit and Jak-2 interacts with the IFN-γR2 subunit of the IFN-γ receptor. The IFN-γR2 chain limits sensitivity to IFN-γ, and the IFN-γR1 chain is usually in excess. But the expression level of IFN-γR2 can be tightly regulated depending on the state of cell differentiation or activation. Receptors are expressed on the surface of almost all cell types. The expression level is determined by the cell type and its activation status. Initially, IFN-γ binds to IFN-γR1, and the formed IFN-γ:IFN-γR1 complex facilitates its binding to IFN-γ-R2, then events of the downstream signaling pathway are initiated [22].

Transcriptional activation of IFN-γ genes occurs through several mechanisms. The most studied response to IFN-γ is mediated by the STAT-1-containing transcription factor GAF (gamma-activated factor), which is activated by the action of tyrosine kinases Jak1 and Jak2 and binds to the GAS (Gamma Activating Sequence), present in the promoter regions of many genes. As a result of gene activation, the formation of a cellular immune response to a viral infection begins [23]. The JAK/STAT pathway is the main signaling pathway initiated by IFN-γ stimulation. Further, IFN-γ, together with one of its receptor subunits IFN-γR1 and pSTAT1, translocates to the cytoplasmic domain in combination with endocytosis and induces gene expression by binding to GAS elements in the promoter region of inducible IFN genes [24].

Activation of receptor-associated JAKs leads to subsequent phosphorylation, activation, and dimerization of latent cytoplasmic STAT transcription factors. The IFN-γ signaling pathway is negatively regulated by SHP-phosphatases (Shp2) or proteins from the cytokine signaling suppressor family (SOCS), mainly SOCS1 and SOCS3 in the cytoplasm, which are involved in the innate and subsequent adaptive immune responses. SOCS-1 binds to Jak1/2, interfering with tyrosine kinase activity and inhibiting further IFN-γ signaling [25]. This pathway can be inhibited by a protein-based inhibitor of activated STATs, which prevents gene transcription by inducing STAT1 dephosphorylation and DNA release [3]. IFN-γ induces genes called interferon-stimulated genes (ISGs) that are both positive and negative regulators of inflammatory signaling [26].

IFN-γ stimulated cells overexpress interferon regulatory factor-1 (IRF-1), a member of the IRF family, which induces the expression of multiple genes involved in biological processes, such as cell cycle regulation, apoptosis, tumor growth inhibition, activation of the synthesis of related molecules, associated with HLA class I, which increases the sensitivity of cells, exposed to IFN-γ, to cytotoxic attacks on T cells [27].

When viruses inhibit the activation of STAT1 molecules, IFN-γ can independently induce a *non-canonical signaling pathway* [28] in which IFN-γ is able to induce gene expression in bone marrow STAT1−/− macrophages, suggesting that IFN-γ acts independently of STAT-1 or in an alternative non-canonical manner. As a rule, the activation of the non-canonical pathway occurs later, after the activation of STAT-1. However, there is evidence that the non-canonical pathway can be activated in the absence or presence of STAT-1 in a dependent manner [29]. The IFN-γ and IFN-α/β signaling pathways intersect at several levels, partially overlapping, which makes it possible for certain functions to cross-talk within the cell. This crossover mechanism is relevant because *in vivo* cells are not stimulated in isolation by a single cytokine, but rather a cytokine cocktail induces gene expression through the integration of multiple signaling pathways.

3. Mechanisms of interferon gamma action

Viruses are intelligent living organisms because they have the ability to invade intracellular organelles and infect host cells [30]. IFN-γ has a direct antiviral effect on infected cells, activates local infiltrating dendritic cells, macrophages and NK cells, modulates the differentiation and maturation of T and B cells, enhances inflammation and antiviral functions [31]. Some of the well-studied antiviral functions of IFN-γ are largely devoid of a specific antiviral mechanism. For example, IFN-γ is a potent inducer of indolamine-2,3-dioxygenase (IDO) and nitric oxide synthase (NOS) [32]. Tryptophan depletion and nitric oxide (NO) production due to IDO and NOS expression, exhibit pronounced antiviral effects, the molecular details of which are mostly unclear [33]. The suppression of any stage of viral life cycle can lead to the inhibition of viral genome replication. IFN-γ is a potent antiviral cytokine that interferes with various stages of the viral life cycle in stimulated cells using the following mechanisms: 1. Inhibits the penetration of virus, both at the extracellular and intracellular stages, by controlling the expression and/or distribution of receptors necessary for the penetration of virus. 2. Inhibits viral replication by disrupting viral replication niche. 3. Disrupts gene expression, preventing translation. 4. Violates stability by interfering with the assembly of the nucleocapsid. 5. Violates the release of virus by breaking the disulfide bond of the required site for cellular interaction. 6. Changes virus reactivation by suppressing the main regulator of viral transcription. 7. Inhibits the penetration of virus at the stage of the invasive viral transfer from endosome to cytoplasm [31]. IFN-γ can also exhibit non-cytolytic antiviral activity against certain viruses. However, the specific targets and effector proteins of the IFN-γ-dependent antiviral response are largely unknown [34].

Tuberculosis (TB) is a chronic infection accompanied by complex changes in both specific and nonspecific reactivity in a patient's body. During TB infection at an early stage (several hours), type I and II interferons are produced as the first line of immune defense in order to attract the largest number of dendritic cells and macrophages to the site of the lesion, that trigger active phagocytosis and inactivate the pathogen. The optimal immune response is formed a few days after infection. Each form of the tuberculous inflammatory process is characterized by an individual pattern of immunological changes. In patients with tuberculomas, there is a decrease in the phagocytic and functional-metabolic activity of monocytes and NK cells, an increase in the number of CD11b + and CD11c + adhesion molecules on granulocytes, and the number of T-lymphocytes. Infiltrative tuberculosis is accompanied by an increase in the population of monocytes with intensive expression of HLA-DR on them, granulocytes are characterized by the growth of the expression of CD11b + and CD11c + adhesion molecules, the number of T-lymphocytes falls [35]. CD4 + Th1 cells and macrophages play the main regulatory role in the development of the immune response in TB. Quantitative and qualitative imbalance of regulatory subpopulations of T-lymphocytes is accompanied by interleukin-dependent immune disorders and other pronounced changes in the cytokine system directing the TB process along a productive or exudative, caseous pathway. CD4+ and CD8+ effector T-cells are sent to the affected area, and begin to induce type II interferons (IFN-γ) greatly shifting the balance towards this class of cytokines and reducing the risk of developing an active TB process. These T-cells, by producing IFN-γ, increase the efficiency of infiltrated phagocytic cells, especially polymorphonuclear neutrophils [36], by stimulating NO and thereby maintaining local host defense [37].

IFN-γ acts as inhibitor of continuous IL-1β production and recruitment of neutrophils, preventing tissue damage. This adaptive immune response allows suppression of innate inflammatory pathways during the development of persistent TB infection. IFN-γ and IFN-γ-dependent NO plays an extremely important role not only in boosting the resistance to Mtb due to antimicrobial activity, but also in the survival of the macroorganism during this chronic infection [38].

Genetics is also very important for resistance and susceptibility to TB. In a study by Sérgio C. A. et al. the populations of lung dendritic cells derived from genetically different hosts have been studied in terms of their role in the size and function of CD4+ populations. At 30 days after infection with H37Rv *M. Tuberculosis*, C57BL/6 mice, which generate a stronger IFN-γ-mediated immune response than BALB/c mice, showed a higher number of CD11c + CD11b-CD103+ cells in the lungs compared to BALB/c mice that showed a high frequency of CD11c + CD11b + CD103 cells. CD11c + CD11b-CD103+ cells purified from the lungs of C57BL/6 of infected mice induced higher concentrations of CD4+ IFN-γ producing cells. This pattern of immune response seems to be related to the genetic characteristics of the host. The authors of the work concluded that genetic differences can reveal immunological biomarkers for the development of tests that predict the progression of TB infection [39].

Therefore, if there is a defect (genetically determined or not) in the immune mechanism, recruited macrophages can facilitate infection by providing the microorganism with an opportunity for intracellular growth and spread. Type I IFN in this case will contribute to the development of TB disease by inducing IL-10 and deactivating macrophages. The first level of immune protection will be broken and the cellular response to the antigen blocked, as a result a latent form of tuberculosis may develop. At the same time, mycobacteria *tuberculosis* (MBT) can suppress the synthesis of endogenous IFN-γ by secreting zinc metalloprotease (ZmpA) inhibiting the production of IL-1β by the host cell. It suppresses the synthesis of PI(3)P, slows down the maturation of phagosomes, and contributes to the development of tuberculosis disease. During the development of TB infection IFN-γ provides the relationship between the two most important links in the immune response of the macroorganism enhancing the antigen-dependent immune response and stimulating the work of phagocytes. IFN-γ activating macrophages attracts them to the focus of infection, increases their ability to destroy absorbed mycobacteria, induces the release of nitric oxide - there is the only inducer of the synthesis of the MHC class II protein in the cell (APC for extracellular pathogens). To kill the mycobacteria survived in the phagosome under the influence of IFN-γ an autophagosome is formed in the macrophage cytosol, the bilayer membrane of it captures the phagosome with mycobacteria and merges it with the lysosome destroying the MBT by lysosome enzymes. The productive type of inflammation in TB is observed with the predominance of the immune response by cell type. It is characterized by a relatively high level of CD4, CD8 lymphocytes, CD4/CD8 index, adequate production of IL-1β, IL-2, IL-12, TNF-a, IFN-γ [40].

Tumor is the result of a complex mechanism of interaction between genetic and epigenetic changes that leads to a dysregulation of intercellular relationships and intracellular signaling pathways. The heterogeneous cellular composition of the tumor and the microenvironment altered by the tumor both limit the effectiveness of standard chemotherapy due to internal, already existing or acquired drug resistance, as well as due to the suppression of apoptosis [41]. One of the recent studies shows that resistance to immunotherapy with checkpoint inhibitors is attributed to defects in the IFN-γ signal [42].

The role of IFN-γ in modulating immune responses is enormous [43–59]. IFN-γ is considered a key component in the immune control of cancer, stimulation of antitumor immunity, and aiding in the recognition and elimination of tumors [43, 59–64]. In addition to activating APCs, enhancing the expression of a number of cytokines (IL-12 and IL-18) leading to the differentiation of Th-1 cells into cytotoxic lymphocytes, induction of a signal cascade in T cells to ensure their effector functions and activation of the expression of molecules of HLA, that is, the realization of cytotoxicity against the tumor, IFN-γ also causes regression of the vascular system of the tumor. Thus, it is possible that IFN-γ slows down tumor growth by inducing its ischemia [45, 65].

The direct antitumor effect of IFN-γ was revealed in several experimental models, however, the mechanisms for the development of this effect were different. So, colorectal cancer cells, it caused apoptosis associated with autophagy by induction of reactive oxygen species by mitochondria. In the T98G glioblastoma line, the induction of apoptosis is due to the suppression of the PI3K/AKT pathway, while the apoptosis of another glioblastoma cell line (U87MG) occurred independently of the PI3K/AKT signaling pathway, through the activation of NF-kB. In human pancreatic carcinoma cells, IFN-γ induces apoptosis in a caspase-1-dependent manner [56–68]. IFN-γ can induce the activation of some micro-RNAs that have an antitumor effect. Thus, it has been shown on melanoma cell lines that activation of miR-29a/b via STAT-1 by IFN-γ leads to an increase in the rate of IFN-γ with other molecules to implement the antitumor effect [63, 69]. However, the response of myeloid cells and other hematopoietic cells to IFN-γ was insufficient for tumor regression, while the effect of IFN-γ on endothelial cells provided a significant antitumor effect. That is, for the development of the antitumor effect, the action of IFN-γ directly on tumor cells and tumor-infiltrating lymphocytes is not enough, but its effect on stromal cells is also necessary [70–73].

The mechanism of the complex immune response to cancer may depend on tumor microenvironment [74]. Unfortunately, the tumor can avoid exposure to endogenous IFN-γ due to the loss of expression of molecules of HLA I class, due to metabolic stress. In several studies, the loss of MHC I expression in cancer cells correlated with the resistance to checkpoint blockade or adoptive immunotherapy. However, in some cancers with low levels of MHC I, it was possible to increase its expression via exogenous interferon therapy [75–77]. The mechanism of how exogenous IFN restores MHC I expression has not been studied in detail. Thus, the answer to the question of how IFN-γ induces signaling pathways that initiate and propagate the apoptotic cascade remains to be seen [78, 79]. One of studies showed that IFN causes increased histone acetylation, demethylation of DNA promoters of TAP genes and immunoproteasomes [77]. Possibly IFN induces IRF1 [80] or stimulates NK cells, which self-disinhibit the launch of their effector mechanisms through the expression of killer inhibitory receptors (KIRs), when they encounter cells with an abnormally low level of MHC I. This is just one of the possible mechanisms described [79, 81].

One more significant fact is that IFN-γ in combination with TNF-α induces the expression of MUC16, a mucin involved in carcinogenesis in breast, ovarian and endometrioid tumors [82].

4. Clinical importance of interferon gamma

Numerous studies are published on the clinical efficacy of IFN-γ in herpesvirus infections (herpes simplex virus type 1 and 2, varicella-zoster virus, Epstein-Barr virus) [78, 83–87]. The use of IFN-γ has been studied in viral complications after organ transplantation, in purulent-septic diseases of newborns, postnatally acquired cytomegalovirus infection, mumps, multiple sclerosis and various bacterial diseases [88–92]. IFN-γ has been used in the complex treatment of patients with human papillomavirus infection, according to the published study results, decreases in virus titer, improvement in the condition of patients, a decrease in the duration and severity of relapses, and faster clinical recovery of patients [93, 94]. Urological community [95, 96] has shown a positive effect of the use of recombinant human IFN-γ on chronic prostatitis therapy, expressed in a decrease in pain syndrome, difficulty urinating and improving the quality of life of the patient. IFN-γ has also been used to inhibit Ebola virus infection in macrophages, an early cellular target of infection [97]. The successful treatment of persistent urethroprostatitis with the identified association of sexually transmitted infections is described. Positive dynamics of the most important immunological parameters was noted after complex treatment with the use of IFN-γ [98, 99].

The inclusion of recombinant IFN-γ in therapy of influenza contributes to a more rapid relief of catarrhal and respiratory symptoms both in adults and children [100]. It was found that the universal risk factor for the development of complications in influenza in children is a low blood level of IFN-γ both in the acute period and in dynamics [101, 102]. IFN-γ exhibits pronounced antiviral activity against various strains of influenza virus, including avian and swine types. The use of exogenous IFN-γ in inhalation combined with subsequent narrow-band optical radiation in pediatrics for acute bronchitis induced by virus infections, including adenovirus, rhinosyncytial (RS) virus, parainfluenza virus with underlying persistence of Epstein Barr virus (EBV), cytomegalovirus (CMV), *S. aureus, S. pneumoniae* and other microorganisms in the lower respiratory tract helped to avoid bacterial complications. Given the fact that influenza viruses can suppress the production of type 1 IFN, the use of type 2 IFN for the prevention and treatment of influenza is advised. The combined use of two types of interferons (alpha and gamma) for the treatment of influenza has also been shown to be promising [103–106]. A comparative, open, randomized study with COVID-19 patients using IFN-γ in terms of changes in the levels of lactate dehydrogenase and C-reactive protein, blood oxygen saturation and other vital functions in the period of inpatient treatment, as well as survival criterium [107, 108].

A systematic review published by J. Ghanavi et al. in 2021 showed that IFN-γ and its receptor (IFN-γR) play a key role in the formation of immunity against MTB and non-tuberculous mycobacteria [30]. The authors emphasized that there is increasing evidence of IFN-γ's important role in host defense against these intracellular pathogens by activating macrophages. Studies confirm that IFN-γ is an integral part of various antibacterial "defenses", including granuloma formation and phagosome-lysosome fusion leading to the death of intracellular mycobacteria. The absence or deficiency of IFN-γ correlates with the overgrowth of intracellular bacteria and the development of tuberculosis infection with mycobacteriosis. New approaches to the treatment of mycobacterial infections are closely related to cell and gene therapy based on the modulation of IFN-γ and IFN-γR.

Meta-analysis on the impact of recombinant IFN-γ on TB patients performed with a number of randomized controlled clinical trials [109] proved its clinical efficacy including for combination of TB with HIV infection. Statistically significant benefits of treatment with recombinant IFN-γ were shown by the results of sputum conversion and X-ray examination of patients. The pooled relative risk (RR) for conversion was 1.97 (95% CI: 1.20–3.24; $p = 0.008$) after 1 month of treatment, 1.74 (95% CI 1.30–2.34; $p = 0.0002$) after 2 months of treatment, 1.53 (95% CI 1.16–2.01; $p = 0.003$) after 3 months of treatment, 1.57 (95% CI 1.20–2.06; $p = 0.001$) after 6 months of treatment and 1.55 (95% CI 1.17–2.05; $p = 0.002$) at the end of treatment. The pooled RR for radiographic progression was 1.38 (95% CI 1.10–1.17, $p = 0.006$) at the end of treatment. Comprehensive treatment with the use of IFN-γ leads to a significant improvement in the indicators of "sleep-rest", "spirituality", "everyday affairs", a decrease in dependence on drugs and medical care. For intramuscularly administered IFN-γ, the meta-analysis included three studies that showed a significant improvement in sputum conversion rates after 2 months of treatment. A randomized controlled trial with aerosolized and subcutaneously administered IFN-γ found a significant reduction in the symptoms of fever, wheezing and night sweats compared with the control group after 1 month of treatment. Meta-analysis suggests that adjuvant therapy using IFN-γ, especially in aerosol form, is effective for patients with TB. IFN-γ within the complex therapy of respiratory TB can significantly increase the effectiveness of anti-tuberculosis therapy (accelerate the cessation of bacterial excretion and closure of cavities in the lungs), prime the immune system and the quality of life of patients. In addition, IFN-γ aerosol may be particularly useful in preventing the development of mycobacterial infections in HIV-infected patients with significantly reduced CD4 cell counts [110].

The results of experimental studies and clinical trials conducted mainly in patients with multi-drug resistant TB, made it possible to propose IFN-γ not only to shorten the long-term standard chemotherapy regimen, however to prevent the latent TB [111]. A. Fortes et al. in 2005 [112] showed that the patients infected with an antibiotic-resistant MBT strain are characterized by a reduced level of endogenous IFN-γ compared to normal patients, and the additional exposure to exogenous IFN-γ in the first months of treatment may lead to the induction of immune system. In this case the appointment of exogenous IFN-γ becomes, in fact, a replacement therapy that can compensate for the endogenous deficiency of the cytokine.

A lot of data has been accumulated regarding the role of IFN-γ in tumor therapy. Antitumor activity of exogenous IFN-γ seems promising for the subsequent development of immunotherapeutic strategies for the complete eradication of cancer. At the same time, it is critical also for the further use of interferon drugs in cancer patients. Clinical studies have shown the effectiveness of IFN-γ therapy in combination with cyclophosphamide and cisplatin, which provided a significant increase in progression-free survival in ovarian cancer. Thus, in a randomized controlled trial 148 patients undergoing primary surgery for stage IC-IIIC ovarian cancer received subcutaneous IFN-γ. In the control group, women received 100 mg/m2 cisplatin and 600 mg/m2 cyclophosphamide, the experimental group included the above regimen with IFN-γ 0.1 mg subcutaneously on days 1, 3, 5, 15, 17, and 19 every 28-day cycle. Progression-free survival (PFS) at 3 years improved from 38% in the control group to 51% in the treatment group, corresponding to median progression times of 17 and 48 months (P = 0.031, relative risk of progression 0.48, CI 0.28–0.82). Overall three-year survival was 58% and 74%, respectively (not significant, median not yet reached). Complete clinical responses were observed in 68% with IFN-γ compared to 56% in controls (not significant). Toxicity was comparable in both groups, with the exception of a mild flu-like syndrome, which was observed in most patients after administration of IFN-γ. Thus, with acceptable toxicity, the inclusion of IFN-γ in first-line ovarian cancer chemotherapy has the advantage of prolonging progression-free survival. This study showed that IFN-γ in combination with carboplatin and paclitaxel is safe as a first-line treatment in patients with advanced ovarian cancer [113]. In an early study by Pujade-Lauraine et al. [114], human recombinant IFN-γ was administered intraperitoneally to patients with stage IIb, IIc, III epithelial ovarian cancer when peritoneal involvement was detected at laparotomy. The study involved 108 patients who received IFN-γ at a dose of 20 × 10 IU/m2 intraperitoneally twice a week for 3–4 months in the absence of clinical manifestations of the disease. IFN-γ response was assessed by exploratory laparotomy. Of 98 patients, 31 (32%) achieved a surgically confirmed response, including 23 patients (23%) with a complete response (CR). Significant prognostic factors for response to IFN-γ were age and size of the residual tumor: a CR rate of 41% was observed in 41 patients younger than 60 years of age and with a residual tumor size of less than 2 cm. The probability of response was independent of previous response to first-line chemotherapy. The median duration of response was 20 months and the 3-year survival rate was 62%. IFN-γ response was the most significant predictor of survival in patients with residual disease. Side effects included fever, flu-like symptoms, neutropenia, and abnormal liver enzyme levels. No significant peritoneal fibrosis was noted. Thus, this work conclude that intraperitoneal administration of IFN-γ promote antitumor response in ovarian cancer [114]. In the study performed by Schmeler et al. [115], human recombinant IFN-γ was administered as subcutaneous injection before and after intravenous carboplatin to patients with recurrent, platinum-sensitive ovarian, fallopian tube and primary peritoneal cancer. The study enrolled 59 patients who received IFN-γ at a fixed dose of 100 mcg on the fifth and seventh day of each 7-day cycle of GM-CSF. IFN-γ response was assessed using the modified World Health Organization Response Evaluation Criteria in Solid Tumors (RECIST). Of the 54 evaluable patients, 9 (17%) achieved a complete response, 21 patients (39%) with a partial response. The overall response rate was 56%. No patients showed treatment-related deaths [115]. Marth et al. [116] in a phase I/II trial tested whether IFN-γ was safe to use it in combination with current standard of care, paclitaxel and carboplatin, in patients with ovarian cancer. Thirty-four patients with newly diagnosed advanced stage III/IV epithelial ovarian cancer were treated with six to nine cycles of paclitaxel (175 mg/m2) and carboplatin ([AUC] 5) every 3 weeks. IFN-γ was administered in increasing doses from 6 days/cycle 0.025 mg SC to 9 days/cycle 0.1 mg SC. As expected, IFN-γ administration was associated with flu-like symptoms. Grade 3/4 neutropenia was observed in 74% (25 of 34) of patients. Other side effects, in particular peripheral neuropathies, were within the previously observed ranges for the combination of paclitaxel + carboplatin. The overall response rate in patients who received either six or nine doses (0.1 mg) of IFN-γ/cycle (n = 28) was 71%. Thus, this study demonstrated the safety of using IFN-γ in combination with carboplatin and paclitaxel for the first-line treatment of patients with ovarian cancer [116].

Intravesical IFN-γ instillations in bladder cancer have been shown to be effective in preventing recurrence. The study included 123 patients with stage Ta, T1, grade 2 tumors who underwent transurethral tumor resection. In group A, 60 patients received IFN-γ (1.5 × 10(7) IU/instillation), while 63 patients from control group B received mitomycin C (40 mg/instillation). During the year of therapy, the following regimen was used: 8 weeks weekly, then four times every two weeks, and then eight monthly instillations for both regimens. The immunophenotypes of intratumoral and intramural leukocytes were also analyzed by immunohistochemical methods and using flow cytometry. As a result of the treatment, relapse was not observed in group A in 44 of 60 (73.4%) patients and in group B in 36 of 63 (57.2%) during a mean follow-up period of 26.5 months (range 3–49 months). After IFN-γ instillations, tissue samples and bladder washes showed a statistically significant increase in the number of T cells: T-helpers, T-cytotoxic cells, natural killer cells and total leukocytes, as well as the number of B cells expressing MHC I, and total leukocytes expressing HLA-DR [117].

Effects of IFN-γ during the adjuvant treatment of radically operated patients with lung adenocarcinoma were evaluated by Pyltsin SP et al. [118] according to the dynamics of the immune status. The study enrolled 63 patients with morphologically verified stages I-IIIA of lung adenocarcinoma. Radical extended pneumonectomy was performed in 17 (26.9%) patients, extended lobectomy - in 42 (66.7%), sublobar resections - in 4 (7.9%). Radically operated patients were randomized into two groups, comparable in terms of the main anthropometric and clinical criteria. By the 21st day after the operation, there were no significant differences in the parameters of cellular immunity in patients of the compared groups, however, significant differences were observed when comparing the parameters of radically operated patients and healthy individuals. Obviously, the transient nature of secondary induced immunosuppression in a tumor process was at least of a prolonged nature (up to 3 months or more) possibly due to the surgical intervention. However, the lack of a trend towards normalization of indicators suggests an increase in immune deficiency, provided by sufficiently aggressive and prolonged adjuvant cytotoxic therapy. After the 1st course, there were no statistically significant differences between the groups, the indicators of both the main and control groups remain lower than in healthy individuals. The changes detected in both groups demonstrated, as before the start of adjuvant treatment, a certain lack of cellular immunity in the form of suppression of T-helpers (CD3 + CD4+) and depression of natural killers (CD56+). The study of the further dynamics of the state of the cellular link of immunity showed that, as a result of IFNγ use after the second course of adjuvant drug therapy, statistically significant differences were detected. In the study group, compared with the control group, the number of T-helper lymphocytes significantly increased (34.1 ± 0.7% versus 31.8 ± 0.8%; $p < 0.05$). The opposite dynamics was observed in relation to cytotoxic T-lymphocytes, the level of which in the study group got statistically significantly lower compared with the control group (26.2 ± 0.6% and 29.8 ± 0.8%, respectively, $p < 0.05$). At the same time, there was found a statistically significant increase in the ratio of CD4+ to CD8+, equal to 1.21 ± 0.04 in the study group compared with 1.04 ± 0.016 in the control group ($p < 0.05$). Depression of NK persisted, both in the study and in the control group, against the background of an increase in the relative number of cytotoxic T cells (CD8+ lymphocytes), especially in the control group. Such changes can develop as a result of compensation for the reduced functional activity of NK by cytotoxic T-lymphocytes. Thus, the administration of exogenous IFNγ leaded to favorable dynamics of the T-helper-inductor link in the patients of the study group, that suggests the immunomodulatory effect of IFNγ on immunocompetent cells with the CD4+ phenotype, which are the main producers of this cytokine when they are differentiated by the Th1 type. Conducting three courses of chemoimmunotherapy to patients of the study group caused the most significant increase in the number of CD4+ lymphocytes, that reached 36.6 ± 0.5% compared with the control group (30.6 ± 0.7%, $p < 0.05$). The opposite dynamics was noted in the content of CD8+ cells, the level of which gradually decreased in the patients of the study group and increased in the control group; at the end of the 3rd course, it was 25.2 ± 0.6% and 31.9 ± 0.6%, respectively ($p < 0.05$). All these changes occurred against the background of a statistically significant increase in the level of T-lymphocytes in the study group (54.5 ± 0.7% vs. 52.5 ± 0.7% in the control group; $p < 0.05$). Conducting adjuvant chemoimmunotherapy with the use of recombinant IFNγ made it possible to achieve a stable correction of the immune status of patients, characterized by the normalization of the main indicators of cellular immunity, with the exception of persistent suppression of the activity of CD56+ cells - NK, effectors of innate immunity, which was also constantly observed during polychemotherapy. In the opinion of the authors, the most important manifestation of the immune action of IFNγ was the leveling of suppression of CD4+ lymphocytes. The full functioning of subpopulations of T-lymphocytes of helpers ensures the regulation of the adaptive cellular immune response, which is very necessary for effective control of the micrometastatic phase of a tumor disease [118].

Early work by Tamura et al. [119] describes the use of recombinant IFN-γ in the treatment of T-cell leukemia in adults. The study involved 5 patients. The drug was administered intramuscularly or intravenously in increasing dosage from 1x10(6) to 8x10(6) JRU (Japan Reference Unit) per day. As a result of the therapy, 1 patient had a complete response, 2 had a partial response, the disease continued to progress in 1 patient, and 1 patient died during the study from pneumonia [119].

The published experience of local application of IFN-γ for the treatment of melanoma of the conjunctiva is represented by several successful clinical cases. The patients received combined treatment with mitomycin C, subconjunctival injections of IFN-γ, and brachytherapy with strontium ophthalmic applicators. IFN-γ at a dose of 500,000 IU was administered under local instillation anesthesia directly under the tumor daily for 10 days. After tumor reduction, brachytherapy was performed. In the first clinical case, after 4 months, a residual radiation reaction was observed in the form of conjunctival hyperemia. Melanoma nodes regressed, areas of flat melanosis remained without progression. In the second clinical case, after combined treatment, a complete regression of the tumor was observed with a follow-up period of 8 months. In the third clinical case, with the follow-up period of 64 months after the first treatment in the area of the tumor cicatricial-modified conjunctiva was found. On the mucosa of the upper eyelid in the center of the scar, there was residual avascular, poorly pigmented tissue. Of the adverse reactions, the researchers observed only local pain at the site of subconjunctival injection of the drug, hyperemia and slight swelling of the conjunctiva and eyelid skin. These side effects did not lead to the continuation of treatment. Thus, the use of IFN-γ can expand the possibilities of organ-sparing treatment of conjunctival melanoma [120].

Clinical experience of IFN-γ application in the treatment of radiation cystitis accompanied with hematuria described by Kaprin AD et al. [121] concerned a study group of 12 patients (vs. 12 in a control group) with late radiation complications from the lower urinary tract. The drug was administered at a dose of 500,000 IU subcutaneously once a day every other day for 20 days (10 injections in total). The effectiveness of the treatment was assessed based on the dynamics of IPSS data (International Prostatic Symptom Score), the degree of pain syndrome, the time of relief of hematuria, and the restoration of urine sterility. The use of IFN-γ made it possible to increase the effectiveness of anti-inflammatory treatment of patients with radiation cystitis: urine sterility was restored in 65% of cases; 4.5 days earlier than in the control group [121].

S.N. Kazakova et al. [122] reported a clinical case of successful rehabilitation approach with local IFN-γ (Ingaron®) use combined to physical therapy in women with postradiation complications as a result of prior endometrial cancer [122].

The summary of antitumor evidence on the clinical efficacy of IFN-γ approaches in oncology is presented in **Table 1**. Current studies are primarily focused on the effects of IFN-γ in oncogynecology, breast cancer and solid tumors.

Undoubtedly, IFN-γ can have direct antiviral and antitumor effect, besides immunomodulating one. However, for a wide clinical use of IFN-y, it is necessary to carry out further experimental research of the mechanisms and conditions that will provide full clinical value and reveal yet hidden potential of this natural pleiotropic molecule for successful health implementation.

Cancer type	Therapy conditions	IFN-γ introduction	Results
Ovarian	First-line	Subcutaneous injection	148 women enrolled. CR were observed in 68%. The median duration of response was 48 months (p = 0.031, RR of progression 0.48, CI 0.28–0.82) and the 3-year survival rate was 74% (n.s., median not yet reached). [113]
Ovarian	Second-line	Intraperitoneal injection	108 patients enrolled. 32% achieved a surgically documented response, including 23% of patients with a CR. A 41% CR rate was observed in patients younger than 60 y.o. and with residual tumor less than 2 cm. The median duration of response was 20 months and the 3-year survival rate in responders was 62%. [114]
Ovarian, fallopian tube and primary peritoneal	Before and after intravenous carboplatin	Subcutaneous injection	59 patients enrolled. Overall response rate was 56% (95% CI: 41–69%) with median time to progression of 6 months. [115]
Ovarian	First-line	Subcutaneous injection	28 patients enrolled. Overall response rate (CR or PR) was 71%. [116]
Bladder	Prevention of recurrence	Intravesical instillation	123 patients enrolled. CR was 73.4% during the median follow-up period of 26.5 months. [117]
Lung adenocarcinoma	Post-surgery adjuvant therapy	Intravenous injection	63 patients enrolled. Improvement of 3-year event-free survival rate by 19 percent (p = 0.06607). [118]
Leukemia	Adjuvant therapy	Intravenous infusion	11 patients enrolled. Overall response rate (CR or PR) was 60% during the median follow-up period of 4 months. [119]
Melanoma of conjunctiva	Local therapy	Subconjunctival injection	3 patients enrolled. Overall response rate (CR or PR) was 60% during the median follow-up period of 5 to 62 months. [120]

CI – confidence interval.
CR – complete response.
PR – partial response.
RR – relative risk.

Table 1.
Summary on clinical evidence of antitumor effects of IFN-γ.

5. Conclusion

IFN-γ occupies a special place in the interferon family. Besides antiviral, it has a strong immunoregulatory effect and plays one of the key roles in the regulation of adaptive immune response in patients with tuberculosis infection, and cancer. Due to its complex action, IFN-γ is quite important for enhancing any viral, mycobacterial and specific tumor clearance.

Author details

Irina A. Rakityanskaya[1], Tat'jana S. Ryabova[2], Anastasija A. Kalashnikova[3], Goar S. Balasaniants[4], Andrej D. Kaprin[5], Feliks I. Ershov[6], Vera V. Kir'janova[7], Tat'jana B. Korzhenevskaja[8], Denis V. Barbinov[9], Andrej V. Ignatovskij[10], Ljudmila Y. Grivtsova[11], Valentina G. Isaeva[12], Natal'ja A. Falaleeva[13], Alisa I. Gil'[14], Svetlana A. Berns[15]*, Natal'ja V. Vasil'eva[16], Julija V. Dolgo-Saburova[17], Elena V. Shagdileeva[18], Ekaterina V. Frolova[19] and Nadezhda S. Astanina[20]

1 Department of Allergology-Immunology and Clinical Transfusiology City Ambulant Department N112, St. Petersburg, Russian Federation

2 Department of Nephrology and Efferent Therapy of the Military Medical Academy Named After S.M. Kirov, St. Petersburg, Russian Federation

3 Laboratory of Clinical Immunology of the FGBU (Federal State Budgetary Institution), A.M. Nikiforov All-Russian Center for Emergency and Radiation Medicine, EMERCOM of Russia (Ministry of Emergency Situations of Russia), St. Petersburg, Russia Federation

4 Department of Phthisiology of the S.M. Kirov Military Medical Academy, St. Petersburg, Russian Federation

5 Ministry of Health of the Russian Federation, National Medical Research Radiological Centre of the Ministry of Health of the Russian Federation, Moscow, Russian Federation

6 National Research Center for Epidemiology and Microbiology Named After the Honorary Academician N.F. Gamaleya, Moscow, Russian Federation

7 Department of Physiotherapy and Medical Rehabilitation of the North-Western State Medical University Named After I.I. Mechnikov (NWSMU), St. Petersburg, Russian Federation

8 FSBI CSC CID of FMBA of Russia, St. Petersburg, Russian Federation

9 SM-Clinic, St. Petersburg, Russian Federation

10 Department of Infectious Diseases, Epidemiology and Dermatovenereology of the St Petersburg University, St. Petersburg, Russian Federation

11 Department of Clinical Immunology of A. Tsyb Medical Radiological Research Center – branch of the National Medical Research Radiological Center of the Ministry of Health of the Russian Federation (A. Tsyb MRRC), Obninsk, Russian Federation

12 Laboratory of Clinical Immunology of A. Tsyb Medical Radiological Research Center – branch of the National Medical Research Radiological Center of the Ministry of Health of the Russian Federation (A. Tsyb MRRC), Obninsk, Russian Federation

13 Department of Drug Treatment of A. Tsyb Medical Radiological Research Center – branch of the National Medical Research Radiological Center of the Ministry of Health of the Russian Federation (A. Tsyb MRRC), Obninsk, Russian Federation

14 The Federal State Budgetary Institution "North-Western District Scientific and Clinical Center Named After L.G. Sokolov Federal Medical and Biological Agency", St. Petersburg, Russian Federation

15 Federal state budgetary institution National Medical Research Center for Therapy and Preventive Medicine of the Ministry of Healthcare of the Russian Federation, Moscow, Russian Federation

16 Research Institute of Medical Mycology named after P.N. Kashkin Federal State Budgetary Educational Institution of Higher Education of the North-Western State Medical University Named After I.I. Mechnikov (NWSMU), St. Petersburg, Russian Federation

17 Research Institute of Medical Mycology Named After P.N. Kashkin Federal State Budgetary Educational Institution of Higher Education of the North-Western State Medical University Named After I.I. Mechnikov (NWSMU), Department of Obstetrics and Gynecology of the Almazov National Medical Research Centre, St. Petersburg, Russian Federation

18 Department of Clinical Mycology, Allergology and Immunology of the North-Western State Medical University Named After I.I. Mechnikov (NWSMU), St. Petersburg, Russian Federation

19 Laboratory of Immunology and Allergology of the Research Institute of Medical Mycology Named After P.N. Kashkin Federal State Budgetary Educational Institution of Higher Education of the North-Western State Medical University named after I.I. Mechnikov (NWSMU), St. Petersburg, Russian Federation

20 Refnot-Pharm Ltd, St. Petersburg, Russian Federation

*Address all correspondence to: svberns@yandex.ru

References

[1] Wheelock EF. Interferon-like virus-inhibitor induced in human leukocytes by phytohemagglutinin. Science. 1965;**149**(3681):310-311. DOI: 0.1126/science.149.3681.310

[2] Zaidi MR, Merlino G. The two faces of interferon-g in cancer. Clinical Cancer Research. 2011;**17**(19):1-7

[3] Alspach E, Lussier DM, Schreiber RD. Interferon γ and its important roles in promoting and inhibiting spontaneous and therapeutic cancer immunity. Cold Spring Harbor Perspectives in Biology. 2018;**11**(3):1-22. DOI: 10.1101/cshperspect.a028480

[4] Lilkova E, Petkov P, Ilieva N, Krachmarova E, Nacheva G, Litov L. Molecular modeling of the effects of glycosylation on the structure and dynamics of human interferon-gamma. Journal of Molecular Modeling. 2019;**25**(127):1-13. DOI: 10.1007/s00894-019-4013-8

[5] Gordon-Alonso M, Hirsch T, Wildmann C, Van Der Bruggen P. Galectin-3 captures interferon-gamma in the tumor matrix reducing chemokine gradient production and T-cell tumor infiltration. Nature Communications. 2017;**8**(793):1-15. DOI: 10.1038/s41467-017-00925-6

[6] Burke JD, Young HA. IFN- γ : A cytokine at the right time, is in the right place. Seminars in Immunology. 2019;**43**:1-8. DOI: 10.1016/j.smim.2019.05.002

[7] Hill N, Sarvetnick N. Cytokines: Promoters and dampeners of autoimmunity. Current Opinion in Immunology. 2002;**14**(6):791-797. DOI: 10.1016/s0952-7915(02)00403-x

[8] Mendoza JL, Escalante NK, Jude KM, Bellon JS, Su L, Horton TM, et al. Structure of the IFNγ receptor complex guides design of biased agonists. Nature. 2019;**567**:56-60. DOI: 10.1038/s41586-019-0988-7

[9] Gattoni A, Parlato A, Vangieri B, Bresciani M, Derna R. Interferon-gamma: Biologic functions and HCV therapy (type I/II) (1 of 2 parts). La Clinica Terapeutica. 2006;**157**(4):377-386

[10] Xu H. Th1 cytokine-based immunotherapy for cancer. Hepatobiliary & Pancreatic Diseases International. 2014;**13**(5):482-494. DOI: 10.1016/s1499-3872(14)60305-2

[11] Marie T, Amos K, Mlecnik B, Fredriksen T, Mauger S, Bindea G, et al. Clinical impact of different classes of infiltrating T cytotoxic and helper cells (Th1, th2, treg, th17) in patients with colorectal cancer. Cancer Research. 2011;**71**(4):1263-1271. DOI: 10.1158/0008-5472.CAN-10-2907

[12] Chaix J, Tessmer MS, Hoebe K, et al. Cutting edge: Priming of NK cells by IL-18. Journal of Immunology. 2008;**181**(3):1627-1631. DOI: 10.4049/jimmunol.181.3.1627

[13] Pien GC, Satoskar AR, Takeda K, Akira S, Biron CA. Cutting edge: Selective IL-18 requirements for induction of compartmental IFN-gamma responses during viral infection. Journal of Immunology. 2000;**165**(9):4787-4791. DOI: 10.4049/jimmunol.165.9.4787

[14] Stern AS, Podlaski FJ, Hulmes JD, Pan YC, Quinn PM, Wolitzky AG, et al. Purification to homogeneity and partial characterization of cytotoxic lymphocyte maturation factor from human

B-lymphoblastoid cells. Proceedings of the National Academy of Sciences of the United States of America. 1990;**87**:6808-6812. DOI: 10.1073/pnas.87.17.6808

[15] Kallal LE, Biron CA. Changing partners at the dance: Variations in STAT concentrations for shaping cytokine function and immune responses to viral infections. Jak-Stat. 2013;**2**(1):e23504. DOI: 10.4161/jkst.23504

[16] Durali D, de Goër de Herve M-G, Giron-Michel J, Azzarone B, Delfraissy J-F, Taoufik Y, In human B cells, IL-12 triggers a cascade of molecular events similar to Th1 commitment. Blood. 2003;**102**:4084-4089. DOI: 10.1182/blood-2003-02-0518

[17] Vignali DAA, Kuchroo VK. IL-12 family cytokines: Immunological playmakers. Nature Immunology. 2012;**13**:722-728. DOI: 10.1038/ni.2366

[18] Hamza T, Barnett JB, Li B. Interleukin 12 a key immunoregulatory cytokine in infection applications. International Journal of Molecular Sciences. 2010;**11**(3):789-806. DOI: 10.3390/ijms11030789

[19] Schoenborn JR, Wilson CB. Regulation of interferon-gamma during innate and adaptive immune responses. Advances in Immunology. 2007;**96**:41-101. DOI: 10.1016/S0065-2776(07)96002-2

[20] Arbouzova NI, Zeidler MP. JAK/STAT signal ling in drosophila: Insights into conserved regulatory and cellular functions. Development. 2006;**133**:2605-2616. DOI: 10.1242/dev.02411

[21] Bach EA, Aguet M, Schreiber RD. The IFN gamma receptor: A paradigm for cytokine receptor signaling. Annual Review of Immunology. 1997;**15**:563-591. DOI: 10.1146/annurev.immunol.15.1.563

[22] Schroder K, Hertzog PJ, Ravasi T, Hume DA. Interferon-gamma: An overview of signals, mechanisms and functions. Journal of Leukocyte Biology. 2004;**75**:163-189. DOI: 10.1189/jlb.0603252

[23] Saha B, Jyothi PS, Chandrasekar B, Nandi D. Gene modulation and immunoregulatory roles of interferon gamma. Cytokine. 2010;**50**(1):1-14. DOI: 10.1016/j.cyto.2009.11.021

[24] Randall RE, Stephen G. Interferons and viruses: An interplay between induction, signalling, antiviral responses and virus countermeasures. Journal of General Virology. 2008;**89**:1-47. DOI: 10.1099/vir.0.83391-0

[25] Song MM, Shuai K. The suppressor of cytokine signaling (SOCS) 1 and SOCS3 but not SOCS2 proteins inhibit interferon-mediatedantiviral and antiproliferative activities. The Journal of Biological Chemistry. 1998;**273**(52):35056-35062. DOI: 10.1074/jbc.273.52.35056

[26] Liu H, Golji J, Brodeur LK, Chung FS, Chen JT, Rosalie S, et al. Tumor-derived IFN triggers chronic pathway agonism and sensitivity to ADAR loss. Nature Medicine. 2019;**25**:95-102. DOI: 10.1038/s41591-018-0302-5

[27] Yang M, Du Q, Varley PR, Goswami J, Liang Z, Wang R, et al. Interferon regulatory factor 1 priming of tumour-derived exosomes enhances the antitumour immune response. British Journal of Cancer. 2018;**118**(1):62-71. DOI: 10.1038/bjc.2017.389

[28] Garcia-Diaz A, Shin DS, Moreno BH, Saco J, Escuin-Ordinas H, Rodriguez GA, et al. Interferon receptor signaling pathways regulating PD-L1 and PD-L2 expression. Cell Reports. 2017;**19**:1189-1201. DOI: 10.1016/j.celrep.2017.04.031

[29] Green DS, Young HA, Valencia JC. Current prospects of type II interferon γ signaling and autoimmunity. The Journal of Biological Chemistry. 2017;**292**(34):13925-13933. DOI: 10.1074/jbc.R116.774745

[30] Smith AE, Helenius A. How viruses enter animal cells. Science. 2004;**304**(5668):237-242. DOI: 10.1126/science.1094823

[31] Kang S, Brown HM, Hwang S. Direct antiviral mechanisms of interferon-gamma. Immune Network. 2018;**18**(5):e33. DOI: 10.4110/in.2018.18.e33

[32] Karupiah G, Xie QW, Buller RM, Nathan C, Duarte C, MacMicking JD. Inhibition of viral replication by interferon-gamma-induced nitric oxide synthase. Science. 1993;**261**:1445-1448. DOI: 10.1126/science.7690156

[33] Samuel CE. Antiviral actions of interferons. Clinical Microbiology Reviews. 2001;**14**(4):778-809. DOI: 10.1128/CMR.14.4.778-809.2001

[34] Ghanavi J, Farnia P, Farnia P, Velayati AA. The role of interferon-gamma and interferon-gamma receptor in tuberculosis and nontuberculous mycobacterial infections. International Journal of Mycobacteriology. 2021;**10**(4):349-357. DOI: 10.4103/ijmy.ijmy_186_21

[35] Berdyugina OV, Yershova AV. Functional and metabolic features of blood phagocytes at different forms of tubercular inflammatory process of lungs. Medical Immunology (Russia)/Meditsinskaya Immunologiya. 2016;**18**(1):21-32. DOI: 10.15789/1563-0625-2016-1-21-32

[36] Wagner C, Kotsougiani D, Pioch M, Prior B, Wentzensen A, Hänsch GM. T lymphocytes in acute bacterial infection: Increased prevalence of CD11b(+) cells in the peripheral blood and recruitment to the infected site. Immunology. 2008;**125**(4):503-509. DOI: 10.1111/j.1365-2567.2008.02863.x

[37] Xue Q, Yan Y, Zhang R, Xiong H. Regulation of iNOS on immune cells and its role in diseases. International Journal of Molecular Sciences. 2018;**19**(12):3805. DOI: 10.3390/ijms19123805

[38] Mishra BB, Rathinam VA, Martens GW, Martinot AJ, Kornfeld H, Fitzgerald KA, et al. Nitric oxide controls the immunopathology of tuberculosis by inhibiting NLRP3 inflammasome-dependent processing of IL-1β. Nature Immunology. 2013;**14**(1):52-60. DOI: 10.1038/ni.2474

[39] Sérgio CA, Bertolini TB, Gembre AF, Prado RQ, Bonato VL. CD11c(+) CD103(+) cells of mycobacterium tuberculosis-infected C57BL/6 but not of BALB/c mice induce a high frequency of interferon-γ- or interleukin-17-producing CD4(+) cells. Immunology. 2015;**144**(4):574-586. DOI: 10.1111/imm.12411

[40] Tian T, Olson S, Whitacre JM, Harding A. The origins of cancer robustness and evolvability. Integrative Biology (Camb). 2011;**3**:17-30

[41] Manguso RT, Pope HW, Zimmer MD, et al. In vivo CRISPR screening identifies Ptpn2 as a cancer immunotherapy target. Nature. 2017;**547**:413-418

[42] Kasahara T, Hooks JJ, Dougherty SF, Oppenheim JJ. Interleukin 2-mediated immune interferon (IFN-gamma) production by human T cells and T cell subsets. Journal of Immunology. 1983;**130**:1784-1789

[43] Corthay A, Skovseth DK, Lundin KU, Rosjo E, Omholt H, Hofgaard PO, et al. Primary antitumor immune response

mediated by CD4+ T cells. Immunity. 2005;**22**:371-383. DOI: 10.1016/j.immuni.2005.02.003

[44] Matsushita H, Hosoi A, Ueha S, Abe J, Fujieda N, Tomura M, et al. Cytotoxic T lymphocytes block tumor growth both by lytic activity and IFN gamma-dependent cell-cycle arrest. Cancer Immunology Research. 2015;**3**:26-36. DOI: 10.1158/2326-6066.CIR-14-0098

[45] Girardi M, Oppenheim DE, Steele CR, Lewis JM, Glusac E, Filler R, et al. Regulation of cutaneous malignancy by gammadelta T cells. Science. 2001;**294**:605-609. DOI: 10.1126/science.1063916

[46] Gao Y, Yang W, Pan M, Scully E, Girardi M, Augenlicht LH, et al. Gamma delta T cells provide an early source of interferon gamma in tumor immunity. The Journal of Experimental Medicine. 2003;**198**:433-442. DOI: 10.1084/jem.20030584

[47] Ribot JC, deBarros A, Pang DJ, Neves JF, Peperzak V, Roberts SJ, et al. CD27 is a thymic determinant of the balance between interferon-gamma- and interleukin 17-producing gamma-delta T cell subsets. Nature Immunology. 2009;**10**:427-436. DOI: 10.1038/ni.1717

[48] Silva-Santos B, Serre K, Norell H. gammadelta T cells in cancer. Nature Reviews. Immunology. 2015;**15**:683-691. DOI: 10.1038/nri3904

[49] Yu J, Wei M, Becknell B, Trotta R, Liu S, Boyd Z, et al. Pro- and antiinflammatory cytokine signaling: Reciprocal antagonism regulates interferon-gamma production by human natural killer cells. Immunity. 2006;**24**:575-590. DOI: 10.1016/j.immuni.2006.03.016

[50] Keppel MP, Saucier N, Mah AY, Vogel TP, Cooper MA. Activation-specific metabolic requirements for NK cell IFN-gamma production. Journal of Immunology. 2015;**194**:1954-1962. DOI: 10.4049/jimmunol.1402099

[51] Yoshimoto T, Takeda K, Tanaka T, Ohkusu K, Kashiwamura S, Okamura H, et al. IL-12 up-regulates IL-18 receptor B cells: Synergism with IL-18 for IFN-gamma production. Journal of Immunology. 1998;**161**:3400-3407

[52] Barr TA, Brown S, Mastroeni P, Gray D. TLR and B cell receptor signals to B cells differentially program primary and memory Th1 responses to salmonella enterica. Journal of Immunology. 2010;**185**:2783-2789. DOI: 10.4049/jimmunol.1001431

[53] Bao Y, Liu X, Han C, Xu S, Xie B, Zhang Q, et al. Identification of IFN-gamma-producing innate B cells. Cell Research. 2014;**24**:161-176. DOI: 10.1038/cr.2013.155

[54] Ohteki T, Fukao T, Suzue K, Maki C, Ito M, Nakamura M, et al. Interleukin 12-dependent interferon gamma production by CD8 alpha+ lymphoid dendritic cells. The Journal of Experimental Medicine. 1999;**189**:1981-1986. DOI: 10.1084/jem.189.12.1981

[55] Nguyen KB, Cousens LP, Doughty LA, Pien GC, Durbin JE, Biron CA. Interferon alpha/beta-mediated inhibition and promotion of interferon gamma: STAT1 resolves a paradox. Nature Immunology. 2000;**1**:70-76. DOI: 10.1038/76940

[56] Matikainen S, Paananen A, Miettinen M, Kurimoto M, Timonen T, Julkunen I, et al. IFN-alpha and IL-18 synergistically enhance IFN-gamma production in human NK cells: Differential regulation of Stat4 activation and IFN-gamma gene expression by IFN-alpha and IL-12. European Journal

of Immunology. 2001;**31**:2236-2245. DOI: 10.1002/1521-4141(200107)31:7<2236::AID-IMMU2236>3.0.CO;2-G

[57] de Veer MJ, Holko M, Frevel M, Walker E, Der S, Paranjape JM, et al. Functional classification of interferon-stimulated genes identified using microarrays. Journal of Leukocyte Biology. 2001;**69**:912-920. DOI: 10.1189/jlb.69.6

[58] Shankaran V, Ikeda H, Bruce AT, White JM, Swanson PE, Old LJ, et al. IFNgamma and lymphocytes prevent primary tumour development and shape tumour immunogenicity. Nature. 2001;**410**:1107-1111. DOI: 10.1038/35074122

[59] Wang L, Wang Y, Song Z, Chu J, Qu X. Deficiency of interferon-gamma or its receptor promotes colorectal cancer development. Journal of Interferon & Cytokine Research. 2015;**35**:273-280. DOI: 10.1089/jir.2014.0132

[60] Kaplan DH, Shankaran V, Dighe AS, Stockert E, Aguet M, Old LJ, et al. Demonstration of an interferon gamma-dependent tumor surveillance system in immunocompetent mice. Proceedings of the National Academy of Sciences of the United States of America. 1998;**95**:7556-7561. DOI: 10.1073/pnas.95.13.7556

[61] Street SE, Cretney E, Smyth MJ. Perforin and interferon-gamma activities independently control tumor initiation, growth, and metastasis. Blood. 2001;**97**:192-197. DOI: 10.1182/blood.V97.1.192

[62] Street SE, Trapani JA, MacGregor D, Smyth MJ. Suppression of lymphoma and epithelial malignancies effected by interferon gamma. The Journal of Experimental Medicine. 2002;**196**:129-134. DOI: 10.1084/jem.20020063

[63] Enzler T, Gillessen S, Manis JP, Ferguson D, Fleming J, Alt FW, et al. Deficiencies of GM-CSF and interferon gamma link inflammation and cancer. The Journal of Experimental Medicine. 2003;**197**:1213-1219. DOI: 10.1084/jem.20021258

[64] Mitra-Kaushik S, Harding J, Hess J, Schreiber R, Ratner L. Enhanced tumorigenesis in HTLV-1 tax-transgenic mice deficient in interferon-gamma. Blood. 2004;**104**:3305-3311. DOI: 10.1182/blood-2004-01-0266

[65] Wang QS, Shen SQ, Sun HW, Xing ZX, Yang HL. Interferon-gamma induces autophagy-associated apoptosis through induction of cPLA2-dependent mitochondrial ROS generation in colorectal cancer cells. Biochemical and Biophysical Research Communications. 2018;**498**:1058-1065

[66] Zhang R, Banik NL, Ray SK. Combination of all-trans retinoic acid and inter-feron-gamma suppressed PI3K/Akt survival pathway in glioblastoma T98G cells whereas NF-kappaB survival signaling in glioblastoma cells for induction of apoptosis. Neurochemical Research. 2007;**32**:2194-2202. DOI: 10.1007/s11064-007-9417-7

[67] Detjen KM, Farwig K, Welzel M, Wiedenmann B, Rosewicz S. Interferon gamma inhibits growth of human pancreatic carcinoma cells via caspase-1 dependent induction of apoptosis. Gut. 2001;**49**:251-262. DOI: 10.1136/gut.49.2.251

[68] Schmitt MJ, Philippidou D, Reinsbach SE, et al. Interferon-gamma-induced activation of signal transducer and activator of transcription 1 (STAT1) up-regulates the tumor suppressing microRNA-29 family in melanoma cells. Cell Communication and Signaling: CCS. 2012;**10**:41

[69] Braumüller H et al. T-helper-1-cell cytokines drive cancer into senescence. Nature. 2013;**494**:361-365 [PubMed: 23376950]

[70] Zhang B, Karrison T, Rowley DA, Schreiber H. IFNγ - and TNF-dependent bystander eradication of antigen-loss variants in established mouse cancers. The Journal of Clinical Investigation. 2008;**118**:1398-1404 [PubMed: 18317595]

[71] Listopad JJ et al. Fas expression by tumor stroma is required for cancer eradication. Proceedings of the National Academy of Sciences. 2013;**110**:2276-2281 PubMed: 23341634

[72] Kammertoens T, Friese C, Arina A, et al. Tumour ischaemia by interferon-gamma resembles physiological blood vessel regression. Nature. 2017;**545**:98-102

[73] Marijt KA, Sluijter M, Blijleven L, Tolmeijer SH, Scheeren FA, van der Burg SH, et al. Metabolic stress in cancer cells induces immune escape through a PI3K-dependent blockade of IFNγ receptor signaling. Journal for Immunotherapy of Cancer. Jun 13, 2019;7(1):152. DOI: 10.1186/s40425-019-0627-8. PMID: 31196219; PMCID: PMC6567539

[74] Castro F, Cardoso AP, Gonçalves RM, Serre K, Oliveira MJ. Interferon-gamma at the crossroads of tumor immune surveillance or evasion. Frontiers in Immunology. 2018;**9**:847. DOI: 10.3389/fimmu.2018.00847 eCollection 2018

[75] Angell TE, Lechner MG, Jang JK, LoPresti JS, Epstein AL. MHC class I loss is a frequent mechanism of immune escape in papillary thyroid cancer that is reversed by interferon and selumetinib treatment in vitro. Clinical Cancer Research. 2014;**20**:6034-6044. DOI: 10.1158/1078-0432.CCR-14-0879

[76] de Mora-Garcia ML, Duenas-Gonzalez A, Hernandez-Montes J, de la Cruz-Hernandez E, Perez-Cardenas E, Weiss-Steider B, et al. Up-regulation of HLA class-I antigen expression and antigen-specific CTL response in cervical cancer cells by the demethylating agent hydralazine and the histone deacetylase inhibitor valproic acid. Journal of Translational Medicine. 2006;**4**:55. DOI: 10.1186/1479-5876-4-55

[77] van den Elsen PJ, Holling TM, van der Stoep N, Boss JM. DNA methylation and expression of major histocompatibility complex class I and class II transactivator genes in human developmental tumor cells and in T cell malignancies. Clinical Immunology. 2003;**109**:46-52. DOI: 10.1016/S1521-6616(03)00200-6

[78] Shakya AK, O'Callaghan DJ, Kim SK. Interferon gamma inhibits varicella-zoster virus replication in a cell line-dependent manner [published correction appears in J Virol. 2020 mar 17;94(7)]. Journal of Virology. 2019;**93**(12):e00257-e00219. Published 2019 May 29. DOI: 10.1128/JVI.00257-19

[79] Dhatchinamoorthy K, Colbert JD, Rock KL. Cancer immune evasion through loss of MHC class I antigen presentation. Frontiers in Immunology. 2021;**9**(12):636568. DOI: 10.3389/fimmu.2021.636568

[80] Kriegsman BA, Vangala P, Chen BJ, Meraner P, Brass AL, Garber M, et al. Frequent loss of IRF2 in cancers leads to immune evasion through decreased MHC class I antigen presentation and increased PD-L1 expression. Journal of Immunology. 2019;**203**:1999-2010. DOI: 10.4049/jimmunol.1900475

[81] Rosenblatt JD, Podack ER, Barber GN. Advances in Tumor Immunology and Immunotherapy. Augusto Ochoa Springer

Science & Business Media. 2013:371. eBook ISBN 978-1-4614-8809-5

[82] Morgado M, Sutton MN, Simmons M, et al. Tumor necrosis factor-alpha and interferon-gamma stimulate MUC16 (CA125)expression in breast, endometrial and ovarian cancers through NFkappaB. Oncotarget. 2016;7:14871-14884

[83] Milligan GN, Bernstein DI. Interferon-gamma enhances resolution of herpes simplex virus type 2 infection of the murine genital tract. Virology. 1997;**229**(1):259-268. DOI: 10.1006/viro.1997.8441

[84] Andersson J, Isberg B, Christensson B, Veress B, Linde A, Bratel T. Interferon gamma (IFN-gamma) deficiency in generalized Epstein-Barr virus infection with interstitial lymphoid and granulomatous pneumonia, focal cerebral lesions, and genital ulcers: Remission following IFN-gamma substitution therapy. Clinical Infectious Diseases. 1999;**28**(5):1036-1042. DOI: 10.1086/514733

[85] Fujisaki T, Nagafuchi S, Okamura T. Gamma-interferon for severe chronic active Epstein-Barr virus. Annals of Internal Medicine. 1993;**118**(6):474-475

[86] Andersson J. Clinical and immunological considerations in Epstein-Barr virus-associated diseases. Scandinavian Journal of Infectious Diseases. Supplementum. 1996;**100**:72-82

[87] Balachandra K, Thawaranantha D, Ayuthaya PI, Bhumisawasdi J, Shiraki K, Yamanishi K. Effects of human alpha, beta and gamma interferons on varicella zoster virus in vitro. The Southeast Asian Journal of Tropical Medicine and Public Health. 1994;**25**(2):252-257

[88] Antoniou KM, Zervou MI, Tzortzaki EG, Dimadi M, Latsi P, Polychronopoulos V, et al. Old-fashioned colchicine or interferon gamma-1b for the treatment of idiopathic pulmonary fibrosis? The molecular Perspectiv. Chest. 2003;**124**(4_MeetingAbstracts):117S. DOI: 10.1378/chest.124.4_MeetingAbstracts.117S

[89] Ershov FI, Narovlyansky AN. Theoretical and applied aspects of the interferon system: To the 60th anniversary of the discovery of interferons. Problems of Virology. 2018;**63**(1):10-18. (In Russ.). DOI:10.18821/0507-4088-2018-63-1-10-18

[90] Bot A, Bot S, Bona CA. Protective role of gamma interferon during the recall response to influenza virus. Journal of Virology. 1998;**72**(8):6637-6645. DOI: 10.1128/JVI.72.8.6637-6645.1998

[91] Baumgarth N, Kelso A. In vivo blockade of gamma interferon affects the influenza virus-induced humoral and the local cellular immune response in lung tissue. Journal of Virology. 1996;**70**(7):4411-4418. DOI: 10.1128/JVI.70.7.4411-4418.1996

[92] Price GE, Gaszewska-Mastarlarz A, Moskophidis D. The role of alpha/beta and gamma interferons in development of immunity to influenza a virus in mice. Journal of Virology. 2000;**74**(9):3996-4003. DOI: 10.1128/jvi.74.9.3996-4003.2000

[93] Day PM, Thompson CD, Lowy DR, Schiller JT. Interferon gamma prevents infectious entry of human papillomavirus 16 via an L2-dependent mechanism. Journal of Virology. 2017;**91**(10):e00168-e00117. Published 2017 Apr 28. DOI: 10.1128/JVI.00168-17

[94] Rakhmatulina MR, Bolshenko NV. Experience in the use of interferon-gamma in the treatment of sexually transmitted viral infections. Obstetrics and Gynecology. 2018;**12**:102-108. DOI: 10.18565/aig.2018.12

[95] Hrjanin AA, Reshetnikov OV. Interferon-gamma: Horizons of therapy. Antibiotics and Chemotherapy. 2016;**61**(3-4):35-40. Available from: https://cyberleninka.ru/article/n/interferon-gamma-gorizonty-terapii (date of the application: 31.01.2022) [in Russian]

[96] Kaprin AD et al. New abilities in chronic prostatitis treatment. Andrologija i genitalnaja hirurgija. 2010;**4**:36-39

[97] Rhein BA, Powers LS, Rogers K, et al. Interferon-γ inhibits Ebola virus infection. PLoS Pathogens. 2015;**11**(11):e1005263. Published 2015 Nov 12. DOI: 10.1371/journal.ppat.1005263

[98] Molochkov VA, Skirda TA, Aleshkin VA. K voprosu o lechenii persistirujushhego urogenital'nogo hlamidioza. Ross zhurn kozhn i venerich boleznej. 2013;**4**:55-60 [In Russian]

[99] Logvinov LA et al. Current pharmacotherapy of chronic prostatitis. Andrologija i genitalnaja hirurgija. 2012;**3**:49-55

[100] Kir'janova VV et al. Safe comprehensive therapy to prevent complications in children with catarrhal symptoms associated with acute viral respiratory infections: Series of clinical cases. IJPR. 2021;**13**(2):2186-2190. DOI: 10.31838/ijpr/2021.13.02.284

[101] Momot AP, Balackaja IV, Martynenko TI, et al. Indicators of inflammatory reactions, hemostasis and endotheliosis in severe pneumonia depending on the outcomes of the disease and their conjugation. Modern Problems of Science and Education. 2014;**2**:344 [In Russian]

[102] Fesenko OV, Sinopal'nikov AI. Severe community acquired pneumonia and prognosis scales. Practical pulmonology. 2014;**2**:56-59 [in Russian]

[103] Ershov FI. Use of interferons of I and II type in viral infections. Voprosy Virusologii. 2013;(S1):145-154 [In Russian]

[104] Sologub TV, Midikari AS, Agafonov VN, Suzdal'cev AA, Cvetkov VV. Efficiency and expediency of using recombinant gamma interferon in the complex therapy of patients with a(HlNl)pdm09 influenza. Epidemiology and Infectious Diseases. 2017;**22**(2):58-63. DOI: 10.18821/1560-9529-2017-2-58-63 [In Russian]

[105] Shestakova IV, Ivanova MR, Ulyanova YS, Go A, Shabalkin PI. Effectiveness of interferon gamma for the treatment of moderate and severe influenza by neuraminidase inhibitors: The results of an open-label randomized controlled clinical trial (season 2018-2019). Journal of Pulmonary Medicine Respiratory Care. 2020;**2**(1):01-07

[106] Tokin I, Nikiforov V, Shabalkin P, et al. Randomized controlled parallel-design clinical study of the efficacy and safety of intranasal interferon gamma in treatment of influenza-like infections// International Journal of Biomedicine. 2018;**8**(4):327-332. DOI: 10.21103/Article8(4)_OA12

[107] Mjasnikov AL, Berns SA, Ershov FI. Experience in the clinical use of interferon gamma in the complex therapy of patients with coronavirus infection COVID-19. Russian Medical Journal. 2020;**26**(6):394-401. DOI: 10.17816/0869-2106-2020-26-6-394-401 [in Russian]

[108] Mjasnikov AL, Berns SA, Talyzin PA, Ershov FI. Interferon gamma in the treatment of patients with moderate COVID-19. Questions of Virology. 2021;**66**(1):47-54. DOI: 10.36233/0507-4088-24 [in Russian]

[109] Gao XF, Yang ZW, Li J. Adjunctive therapy with interferon-gamma for the

treatment of pulmonary tuberculosis: A systematic review. International Journal of Infectious Diseases. 2011;**15**(9):e594-e600. DOI: 10.1016/j.ijid.2011.05.002

[110] Yola A, Sologub T, Nechaev V, Ivanov A. Immune-based therapy using gamma interferon Ingaron in the treatment of HIV/AIDS patients with active pulmonary tuberculosis (PTB) not previously highly active antiretroviral therapy (HAART). Retrovirology. 2006;**3**(Suppl 1):S38. DOI: 10.1186/1742-4690-3-S1-S38

[111] Reljic R. IFN-gamma therapy of tuberculosis and related infections. Journal of Interferon & Cytokine Research. 2007;**27**(5):353-364. DOI: 10.1089/jir.2006.0103

[112] Fortes KP, Antas PRZ, Franken CLMC, Dalcolmo M, Ribeiro-Carvalho MM, Cunha KS, et al. Sampaio detection of in vitro interferon-gamma and serum tumour necrosis factor-alpha in multidrug-resistant tuberculosis patients. Clinical and Experimental Immunology. 2005;**141**(3):541-548. DOI: 10.1111/j.1365-2249.2005.02872.x

[113] Windbichler GH, Hausmaninger H, Stummvoll W, et al. Interferon-gamma in the first-line therapy of ovarian cancer: A randomized phase III trial. Br J cancer. 2000;82:1138-1144. Journal of Immunotherapy Cancer. 2019;**7**(1):152. DOI: 10.1186/s40425-019-0627-8

[114] Pujade-Lauraine E, Guastalla JP, Colombo N, Devillier P, François E, Fumoleau P, et al. Intraperitoneal recombinant IFN-γ in ovarian cancer patients with residual disease at second-look laparotomy. Journal of Clinical Oncology. 1996;**14**(2):343-350

[115] Schmeler KM et al. A phase II study of GM-CSF and rIFN-γ1b plus carboplatin for the treatment of recurrent, platinum-sensitive ovarian, fallopian tube and primary peritoneal cancer. Gynecologic Oncology. 2009;**113**(2):210-215. DOI: 10.1016/j.ygyno.2009.02.007

[116] Marth C, Windbichler GH, Hausmaninger H, Petru E, Estermann K, Pelzer A, et al. Interferon-gamma in combination with carboplatin and paclitaxel as a safe and effective first-line treatment option for advanced ovarian cancer: Results of a phase I/II study. International Journal of Gynecological Cancer. 2006;**16**(4):1522-1528

[117] Giannopoulos A, Constantinides C, Fokaeas E, et al. The immunomodulating effect of interferon-gamma intravesical instillations in preventing bladder cancer recurrence. Clinical Cancer Research. 2003;**9**:5550-5558

[118] Pyltsin SP, Zlatnik EY, Lazutin YN, Sergostyants GZ, Zakora GI, Leyman IA, et al. Influence of ingaron upon immune state of the patients with adenocarcinoma of lung in the course of adjuvant therapy. MedMedical Immunology. 2014;**16**(6):559-566. DOI:10.15789/1563-0625-2014-6-559-566

[119] Tamura K, Makino S, Araki Y, Imamura T, Seita M. Recombinant interferon beta and gamma in the treatment of adult T-cell leukemia. Cancer. 1987;**59**(6):1059-1062. DOI: 10.1002/1097-0142(19870315)59:6<1059::aid-cncr2820590602>3.0.co;2-m

[120] Brovkina AF, Grishina YY, Yatsenko OY, Andreichenko AM. The first experience with ingaron treatment for conjunctival melanoma. Tumors of the Head and Neck. 2012;**2**:9-12

[121] Kaprin AD et al. Experience of the application of Ingaron (interferon

gamma human recombinant) in the treatment of radiation cystitis accompanied with hematuria. Farmateka. 2013;**12**:40-43

[122] Kazakova SN, Teterina TA, Apolihina IA, Ishhuk MP. An integrated approach to the development of women with post-radiation complications after endometrial cancer (clinical case). Doctor Russia. 2021;**20**(6):97-101. DOI: 10.31550/1727-2378-2021- 20-6-97-101

Section 5

Interferon Gamma and Neurophysiology

Chapter 6

Interferon and HPA Axis: Impact on Neuroimmunological Perturbations

Apoorv Sharma, Abhishek K. Singh, Vijay Kumar and Hridayesh Prakash

Abstract

The interplay between the central nervous system (CNS) and the enteric nervous system (ENS) constitutes the gut-brain axis. This represents a dynamic and bidirectional network of signaling pathways involving the vagus nerve, the immune system, and the molecules released by various microorganisms thriving in our gut. Since humans and bacteria have evolved together and learned to live together in a symbiotic relationship, which is decisive for physio/immune homeostasis of the body. Disruption in this (also known as dysbiosis) is associated with various pathological consequences including several neurological disorders. Out of several pathways that are associated with neurological manifestation, the inflammasome pathway is associated with the progression of multiple sclerosis, Alzheimer's, and Parkinson's disease, depression, schizophrenia, and autism. A growing body of evidence now suggests a reciprocal influence of microbiota and inflammasome activation in the brain. In this chapter, we discuss the cross talk between human gut microbiota and the key immunological signaling processes and their role in CNS development and neurological diseases.

Keywords: interferon, HPA axis, microbiome, immune plasticity and neurological distress

1. Introduction

The gut-brain axis is an integrated system of the central and enteric nervous system and is made up of both neuronal and non-neuronal components of the central nervous system and the peripheral nervous system. The gut-brain axis functions as a directional communication channel and facilitates the interaction of the brain and with the gastrointestinal (GI) system. The complexity of the gut-brain system allows this system to influence a large variety of physiological processes, which include gastric tone alongside, emotions, motivation, and thinking [1]. Gut function analysis is responsible for monitoring and integrating gut functions with emotional and cognitive centers in the brain with peripheral intestinal functions. These communications involve neuro-immuno-endocrine mediators [2].

IntechOpen

The gut-brain axis is the bidirectional communication between the enteric and central nervous systems. The gut microbiota has evolved as a second brain, with interactions between enteric nerves, gastrointestinal tract, and central neural systems leading to important changes in homeostasis [3]. Changes in intestinal microbiota are a major factor in maintaining gut homeostasis, which affects peripheral tissues such as the liver, lungs, spleen, and kidneys. The enteric nervous system communicates with the central nervous system through several pathways including vagus nerve stimulation (VNS), mechanoreceptors from stretch receptors within the intestine that relay sensory information on the mechanical structure of the environment of an organism; hearing (designed for wind movement), smell (for electrical signals) [4]. Enteric neurons also make connections with type II sensory afferents from regions outside the body such as hands' skin or tongue's surface. These connections become prominent during consumption of foods rich in fiber-feces-bacteria interactions, which lead to increased release of gas distending organs and produce similar sensations within intestines/gut via other mechanosensors [5, 6].

2. Gut-brain axis interplay between enteric nervous system and central nervous system

The gastrointestinal system is the main avenue where these two systems interact and amplify endocrine signals that influence behavior quality control, metabolic health coordination involving appetite regulation-autoregulation. Vagus nerve stimulation studies have demonstrated increased excretion zone production and suggested that gastric motility patterns may be modulated by vagus nerve stimuli [7]. Gut microbiome content is now considered a putative indicator of the health of a person. This is well documented with the observation that certain gut microbiota can cure food, drug, and toxin-related sickness. At the cellular level, it mediates the cross talk through immune-system signaling pathways, augmenting the secretion of neurotransmitters such as GABA and subsequent release of endorphins that activate brain receptors [8]. Dysbiotic gut-induced pathogenic inflammation has a direct impact on neurobehavioral functioning, which involves the loss of various neuronal sensing mechanisms, which affect memory performance [2]. The autonomic nervous system is composed of the sympathetic and parasympathetic limbs that help the body regulate the nervous system associated with the gut. It is made up of enteric, spinal, and vagal pathways, which in association with the HPA axis respond and control stress through the efferent axis, which coordinates adaptive responses in the body [9].

The limbic system of the brain is predominantly involved in memory and emotion. Stress and elevated levels of the inflammatory cytokines by HPA activate an opioid system in the brain, leading to the release of cortisol, a stress hormone that connects the brain and the gut reciprocally [10]. The gut microbiome has an important role in these communications as shown in **Figure 1**. The enteric nervous system of the GI tract is derived from a phylum-specific brain known as the enteric central nervous system. This is composed of a sympathetic and inhibitory branch that controls peristalsis and GI motility. In some species, these two divisions are subdivided into separate ganglia [11].

The concerted activity of the neuronal and immune systems keeps the gut healthy by regulating the food we eat and the viruses and bacteria we encounter.

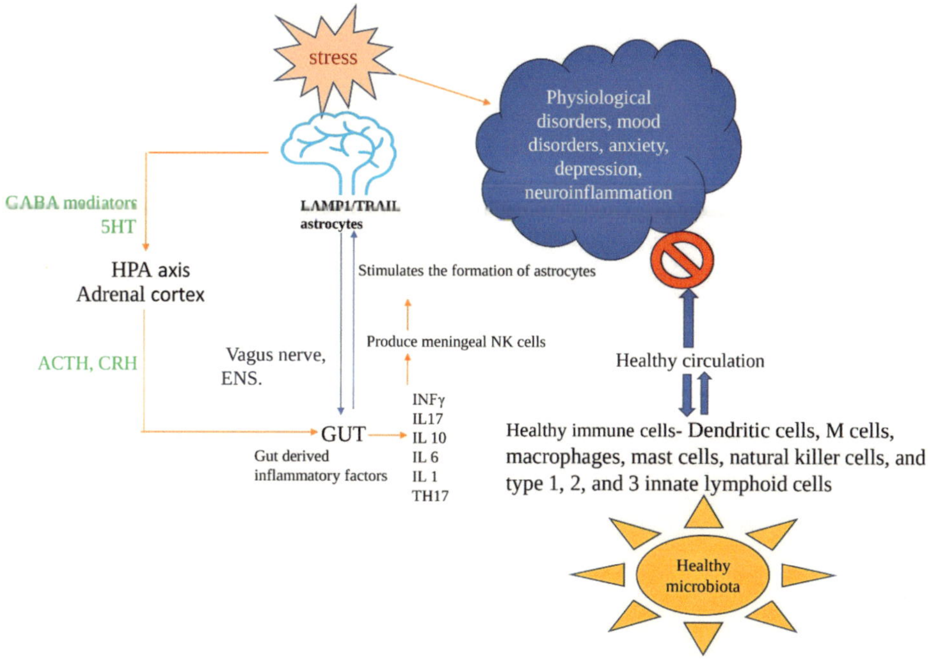

Figure 1.
Gut-brain axis and physiological purturbatin. It shows the role of brain gut-microbiota axis in alleviating the physiological/mood disorders, anxiety, depression, etc., caused due to stress and the interaction between the central nervous system/enteric nervous system (ENS) and gut microbiota. Vagus nerve modulates the gut brain axis. HPA axis; hypothalamus pituitary adrenal Axis, GABA; gamma amino butyric acid, 5HT; 5-hydroxytryptamine, ACTH; adrenocorticotropic-hormone, CRH; corticotropin-releasing hormone.

The enteric nervous system (ENS) senses and reacts to the environment by transmitting chemical signals to the gut cells. This system helps the gut to digest food and fight off infections, which are quite an intricate aspect of the gut-brain axis. Given this, this chapter focuses the mechanism on the neural tubes/connections that maintain gut immune homeostasis. Recent studies on ENS have revealed its importance for microbe-induced immune responses in the gut. It is estimated that the human body hosts approximately 100 trillion microbes, which are essential for proper digestion and metabolic health while others are bad. For instance, bacteria have been associated with gastrointestinal (GI) conditions such as Crohn's disease or colitis. A plethora of evidence revealed the association of gut flora with food allergies including celiac disease. The gut microbiota and their metabolites act as ligands to various cognate receptors that alter epithelial cells' biology as well as their polarity to orchestrate signals from the GI tract to other organs of the body such as the brain or liver [12]. A study found that gut microbiota alters neurons and decreases anxiety levels through neurotransmitter release from neurons into the blood circulation [13].

Recent studies have demonstrated that dysbiosis promotes the prognosis of autism and schizophrenia [14] for which several hypotheses were proposed. Of those, immune metabolic programming of the host seems to be one potential mechanism that predisposes us more susceptible to certain diseases. Alterations in gut microbiota have also been shown to promote obesity, particularly in people with a history of food allergies or an autoimmune disease such as Hashimoto's thyroiditis [15].

3. Inflammatory signaling across the gut-brain axis

The hypothalamus-pituitary–adrenal axis is the main system that controls the production of hormones by the adrenal glands. The hypothalamus controls the production of corticotropin-releasing hormone (CRH), which stimulates the pituitary to produce adrenocorticotropic hormone (ACTH). ACTH then stimulates the production of glucocorticoids by the adrenal glands. Neuro-immunological systems involve interactions between neuronal and immune cells. These systems are responsible for protecting the brain and spinal cord from damage by harmful agents, such as toxins or bacteria. The adrenal hormones cortisol and aldosterone play an important role in this protection. GI tract and the central nervous system (CNS) remain continuously exposed to external and internal antigens. This tweaks intricate cellular networks comprising immune and neural cells, which actively sense harmful stimuli and orchestrate local and systemic bidirectional inflammatory responses. It occurs in both the afferent ("gut-to-brain") and efferent ("brain-to-gut") ways across the gut-brain axis to advocate the host's health status toward homeostasis [16].

GALT (gut-associated lymphoid tissue) responds quickly to intestinal assaults caused by the gut microbiota and/or their primary and secondary metabolites also known as by-products. The first line of defense against intestinal threats is provided by the innate immune cells that are dendritic cells (DCs), M cells, macrophages, mast cells, natural killer (NK) cells, and type 1, 2, and 3 innate lymphoid cells (ILCs). On the other hand, the adaptive immune cells such as CD4+ T effector cells as well as cytotoxic CD8+ T cells [T helper 1 cells (TH1 cells), TH2 cells, and TH17 cells], and regulatory T cells, not only can act directly but also move to distant organs, including the brain [17]. ENS protects the intestinal barrier by secreting RET receptor ligands, which cause ILC3-dependent synthesis of interleukin-22 (IL-22). Through BMP2-BMPR signaling, macrophages engage with enteric neurons to improve neuronal survival and alleviate inflammation-induced bowel dysmotility [18].

There are three primary mechanisms for bidirectional transmission of inflammatory signals between the gut and the CNS. The first one is the humoral pathway that involves the secretion of gut-derived inflammatory factors (IL-1,6, and 17) and 17, IFN γ. These cytokines disrupt BBB integrity [19] and cause developmental abnormalities in the brain. This activates the HPA axis that causes systemic glucocorticoid release, which alters intestinal functions [20]. The second one is the cellular immune pathway in which the intestinal immune cells directly modulate neuro-immune homeostasis and cognitive response toward inflammation. Stress-induced neuroinflammation alters the gut microbiome and releases toxic antigens, which in turn induce maturation of B cells into immunoglobulin A (IgA)-secreting plasma cells, which govern luminal microbial populations. Gut-derived cells may potentially instruct local immune cells in the CNS. A subpopulation of IFNγ-producing meningeal NK cells stimulates the formation of neuro-immunoregulatory astrocytes expressing LAMP-1 and TRAIL [21] as shown in **Figure 1**. Furthermore, the gut microbiota influences the activity of these gut-derived NK cells and their capacity to control astrocytes formation. The third pathway is the neuronal pathway, which is connected with afferent and efferent vagal nerves. The afferent vagal nerve originating from the gut projects into the nucleus solitarius of the brainstem, is endowed with inflammation detecting receptors in the intestines [22]. The impulses are sent to high brain order through the afferent vagal nerve, which triggers the HPA axis. This in turn activates the neural circuits implicated in sickness behavior [23]. Enteric neurons stimulated by cholinergic vagal efferent fibers potentially inhibit the production of inflammatory cytokines IL-1b, IL-6, IL-18, and TNF-a [24] and retune M-1 effector macrophage toward M2 macrophages. These retuned macrophages secrete histamine, 5-HT, enhance voltage-gated channel activity, and promote the excitability of nociceptor neurons [25] for acquiring homeostasis.

4. IFN and intestinal homeostasis

IFN is one of the dual-edge components of the intestinal immune system, which orchestrates intestinal homeostasis and inflammation [26]. Imbalance of the IFN system results in resulting in severe inflammation, cancer, and intestinal damage. A fine balance in the cell proliferation and immunological response is decisive for the GI homeostasis. Type I interferons (IFN-α and β) are pleiotropic cytokines and have both pro and anti-inflammatory manifestations in the gut [26]. Type I IFN counteracts the effects of locally produced IL-17 by blocking the secretion of IL-1, IL-23, osteopontin and increasing the synthesis of IL-27 in DCs [27], which are the main source of Type I IFN [28] and play a crucial role in modulating T-cell-mediated antigen recognition. Type I IFN is known to promote the secretion of anti-inflammatory cytokines (e.g., IL-10, IL-27, and IL-1RA) by triggering the negative feedback PIAS (protein inhibitor of activated STAT) and SOCS (suppressor of cytokine signaling) proteins in phagocytes and T cells [29, 30]. Type I IFN promotes the differentiation of CD4+ Th cells into regulatory T-cells and aids in the maintenance of intestinal homeostasis under constant inflammatory microbial assault during dysbiosis [31].

5. The involvement of gut microbiota in age-related neuro-immune dysbiosis

The gut microbiome is important for healthy aging and long life, and problems with gut dysbiosis may lead to unhealthy aging and shorter lives [32]. Age-related gut dysbiosis tweaks innate immune response and triggers meta-inflammation, which leads to many age-related degenerative pathologies. Disturbance of these communications by age-related dysbiotic gut can affect host health and life span. This also changes the ratio of good v/s bad microbiome, which ultimately dictates host health [33]. With age, gut microbiota becomes more diverse and variable. A disturbed gut microbiome is known to activate the innate immune response and chronic low-grade inflammation, which can lead to many age-related diseases and premature aging. The gut microbiota communicates with the host through various biomolecules, nutrient signaling-independent pathways, and epigenetic mechanisms. These communications can be disrupted by gut dysbiosis in older people, which can affect their health and life span.

Microbiota may be associated with irritable bowel syndrome (IBS) including IBD [34], which has a significant influence on GBA, which is due to enhanced interaction of intestinal cells and ENS locally, but also with CNS peripherally via neuroendocrine and metabolic pathways. One analysis prudently demonstrated the significant improvement and recovery in patients with hepatic encephalopathy after their treatment with antibiotics [35] indicating the significance of gut microbiome on disease management. Other compelling studies support the importance of the gut microbiome in influencing anxiety and depressive-like behaviors [36].

Additional lines of evidence potentially suggest that dysbiotic gut is a common etiological factor of autism [37]. Dysbiosis is common in people with functional gastrointestinal disorders (FGIDs) and is linked with mood disorders [38]. Data show that both brain and gut dysfunctions occur in people with FGIDs, the former being more dominant in people with irritable bowel syndrome (IBS) [39]. This disruption in the gut determines changes in intestinal motility and secretion, causes visceral hypersensitivity, and leads to cellular alterations in the entero-endocrine and immune systems.

Based on several ongoing discussions among the community, we believe that microbiota may be involved in the pathophysiology of various IBS symptoms [34] also. Supplementation of probiotics in conjunctions of antibiotics for curing IBS [40] provides experimental evidence that bad microbiome is directly linked to IBS pathogenesis. [41, 42]. Furthermore, many studies have demonstrated that the microbiota has a role in regulating GBA. Germ-free animals show altered neural function and behavior due to a lack of microbiota colonization [43]. Studies have shown that gut bacterial colonization is important for the development and maturation of the ENS and the CNS [44].

The absence of gut bacteria is associated with alterations in expression and turnover of a neurotransmitter, delayed gastric emptying and intestinal transit, reduced migrating motor complex cyclic recurrence and distal propagation, and enlarged cecal size [45]. Neuromuscular abnormalities resulting from a lack of gut bacteria are restored after animal colonization in a bacterial-species-specific manner. A comparative study between non-germ-free and germ-free (GF) animals has found that the microbiota affects stress reactivity and anxiety-like behavior [46].

The microbiota regulates the thresholds of HPA activities, which is revealed by decreased anxiety of GF animals over non-GF animals toward ACTH and cortisol mediated stress levels supporting the idea that a critical period exists during which the brain is particularly sensitive to the effects of the microbiota. This was further supported by increased memory dysfunction, changes in brain-derived neurotrophic factor (BDNF), alterations in the microbiota, and an increase in serotonin turnover [2] in dysbiotic animals.

Studies have shown that changes in gut microbiota can help in managing anxiety and improve the body's response to stress. These studies also show that manipulating gut microbiota affects brain neurochemistry, with some changes occurring in regions associated with anxiety and stress [47]. This was supported by the observation that probiotics can lower the levels of GABAAα2 mRNA levels in the prefrontal cortex and amygdala but increased them in the hippocampus. This was paralleled by a reduction in stress-induced cortisol release, decreased anxiety and depression-related behavior, and an increase in hippocampal BDNF expression [48]. Several groups have identified that probiotics can increase BDNF levels, reduce age-related changes in the hippocampus, and reverse neonatal maternal separation-induced pain hypersensitivity. These effects are likely due to changes in the microbiota composition of the animals treated with probiotics.

Microbiota communication with the brain involves the vagus nerve, which transmits information from the luminal environment to the CNS. The vagus nerve may play a role in the anxiolytic effect of probiotics. Bacteria communicate with the brain by using the vagus nerve, which transmits information from the surrounding environment to the brain [5]. Studies with animal colitis model caused by anxiety suggest anxiolytic effects of *Bifidobacterium longum* vagotomized animals [48]. Similarly, *Lactobacillus helveticus* R0052 and *B. longum* R0175 prevented changes in hippocampal neurogenesis and expression in hypothalamic genes involved in synaptic plasticity [49].

6. Current and future perspective

The reciprocal exchange of inflammatory signals via the gut-brain axis is critical for the control of physiological activities as well as the pathophysiology of inflammation-related disorders. Intestinal immune cells can be driven to the brain in the afferent direction, but the underlying mechanism remains elusive so far. The gut microbiota has been identified as one of the key regulators of immune cells in the gut-brain axis. Accumulating data suggest that the microbiota communicates with the brain in several ways, including via the vagus nerve, the immune cells, several cytokines, and the luminal environment. Dysbiosis has been linked to several inflammation-related diseases and risk factors such as age, nutrition, and stress. However, it is dubious if dysbiosis has a causative role in these situations. Thus, identifying ways by which certain intestinal microbiota antigens and by-products impact intestinal immune cells or maybe cross the BBB to affect CNS cells would aid in answering dysbiotic-related neuronal disorders.

Author details

Apoorv Sharma[1], Abhishek K. Singh[1], Vijay Kumar[1] and Hridayesh Prakash[1,2*]

1 Amity Institute of Neuropsychology and Neurosciences, Amity University, Noida, India

2 Department of Biotechnology, MM College of Engineering, Maharishi Markandeshwar (Deemed to be University), Mullana, Ambala, Haryana, India

*Address all correspondence to: hprakash@amity.edu; hridayesh.prakash@mmumullana.org

References

[1] Ganci M, Suleyman E, Butt H, Ball M. The role of the brain–gut–microbiota axis in psychology: The importance of considering gut microbiota in the development, perpetuation, and treatment of psychological disorders. Brain and Behavior. 2019;**9**:e01408

[2] Carabotti M, Scirocco A, Maselli MA, Severi C. The gut-brain axis: Interactions between enteric microbiota, central and enteric nervous systems. Annals of Gastroenterology: Quarterly Publication of the Hellenic Society of Gastroenterology. 2015;**28**:203

[3] Rao M, Gershon MD. Enteric nervous system development: What could possibly go wrong? Nature Reviews Neuroscience. 2018;**19**:552-565

[4] Morais LH, Schreiber HL, Mazmanian SK. The gut microbiota–brain axis in behaviour and brain disorders. Nature Reviews Microbiology. 2021;**19**:241-255

[5] Breit S, Kupferberg A, Rogler G, Hasler G. Vagus nerve as modulator of the brain–gut axis in psychiatric and inflammatory disorders. Frontiers in Psychiatry. 2018;**44**

[6] Spencer NJ, Hu H. Enteric nervous system: Sensory transduction, neural circuits and gastrointestinal motility. Nature Reviews Gastroenterology & Hepatology. 2020;**17**:338-351

[7] Lu KH, Cao J, Oleson S, Ward MP, Phillips R, Powley TL, et al. Vagus nerve stimulation promotes gastric emptying by increasing pyloric opening measured with magnetic resonance imaging. Neurogastroenterology & Motility. 2018;**30**:e13380

[8] Moloney RD, Desbonnet L, Clarke G, Dinan TG, Cryan JF. The microbiome: Stress, health and disease. Mammalian Genome. 2014;**25**:49-74

[9] Tsigos C, Chrousos GP. Hypothalamic–pituitary–adrenal axis, neuroendocrine factors and stress. Journal of Psychosomatic Research. 2002;**53**:865-871

[10] Mayer EA, Savidge T, Shulman RJ. Brain–gut microbiome interactions and functional bowel disorders. Gastroenterology. 2014b;**146**:1500-1512

[11] Browning KN, Travagli RA. Central nervous system control of gastrointestinal motility and secretion and modulation of gastrointestinal functions. Comprehensive Physiology. 2014;**4**:1339

[12] Valdes AM, Walter J, Segal E, Spector TD. Role of the gut microbiota in nutrition and health. BMJ. 2018;**361**

[13] Clapp M, Aurora N, Herrera L, Bhatia M, Wilen E, Wakefield S. Gut microbiota's effect on mental health: The gut-brain axis. Clinics and Practice. 2017;**7**:131-136

[14] Agustí A, García-Pardo MP, López-Almela I, Campillo I, Maes M, Romaní-Pérez M, et al. Interplay between the gut-brain axis, obesity and cognitive function. Frontiers in Neuroscience. 2018;**12**:155

[15] Virili C, Fallahi P, Antonelli A, Benvenga S, Centanni M. Gut microbiota and Hashimoto's thyroiditis. Reviews in Endocrine and Metabolic Disorders. 2018;**19**:293-300

[16] Agirman G, Yu KB, Hsiao EY. Signaling inflammation across

the gut-brain axis. Science. 2021a;**374**:1087-1092

[17] Rojas OL, Pröbstel A-K, Porfilio EA, Wang AA, Charabati M, Sun T, et al. Recirculating intestinal IgA-producing cells regulate neuroinflammation via IL-10. Cell. 2019a;**176**:610-624

[18] Huh JR, Veiga-Fernandes H. Neuroimmune circuits in inter-organ communication. Nature Reviews Immunology. 2020a;**20**:217-228

[19] Parker A, Fonseca S, Carding SR. Gut microbes and metabolites as modulators of blood-brain barrier integrity and brain health. Gut Microbes. 2020a;**11**:135-157

[20] Agirman G, Yu KB, Hsiao EY. Signaling inflammation across the gut-brain axis. Science. 2021b;**374**:1087-1092

[21] Sanmarco LM, Wheeler MA, Gutiérrez-Vázquez C, Polonio CM, Linnerbauer M, Pinho-Ribeiro FA, et al. Gut-licensed IFNγ+ NK cells drive LAMP1+ TRAIL+ anti-inflammatory astrocytes. Nature. 2021a;**590**:473-479

[22] Bonaz B, Bazin T, Pellissier S. The vagus nerve at the interface of the microbiota-gut-brain axis. Frontiers in Neuroscience. 2018a;**12**:49

[23] Bonaz B, Bazin T, Pellissier S. The vagus nerve at the interface of the microbiota-gut-brain axis. Frontiers in Neuroscience. 2018b;**12**:49

[24] Bluthe R-M, Michaud B, Kelley KW, Dantzer R. Vagotomy blocks behavioural effects of interleukin-1 injected via the intraperitoneal route but not via other systemic routes. Neuroreport. 1996a;**7**:2823-2827

[25] Yam MF, Loh YC, Tan CS, Khadijah Adam S, Abdul Manan N, Basir R. General pathways of pain sensation and the major neurotransmitters involved in pain regulation. International Journal of Molecular Sciences. 2018a;**19**:2164

[26] Kotredes KP, Thomas B, Gamero AM. The protective role of type I interferons in the gastrointestinal tract. Frontiers in Immunology. 2017;**8**:410

[27] Shinohara ML, Kim J-H, Garcia VA, Cantor H. Engagement of the type I interferon receptor on dendritic cells inhibits T helper 17 cell development: Role of intracellular osteopontin. Immunity. 2008;**29**:68-78

[28] Hoshino K, Kaisho T, Iwabe T, Takeuchi O, Akira S. Differential involvement of IFN-β in toll-like receptor-stimulated dendritic cell activation. International Immunology. 2002;**14**:1225-1231

[29] Guarda G, Braun M, Staehli F, Tardivel A, Mattmann C, Förster I, et al. Type I interferon inhibits interleukin-1 production and inflammasome activation. Immunity. 2011;**34**:213-223

[30] Wang Q, Miyakawa Y, Fox N, Kaushansky K. Interferon-α directly represses megakaryopoiesis by inhibiting thrombopoietin-induced signaling through induction of SOCS-1. Blood, The Journal of the American Society of Hematology. 2000;**96**:2093-2099

[31] Lee SE, Li X, Kim JC, Lee J, González-Navajas JM, Hong SH, et al. Type I interferons maintain Foxp3 expression and T-regulatory cell functions under inflammatory conditions in mice. Gastroenterology. 2012;**143**:145-154

[32] Ragonnaud E, Biragyn A. Gut microbiota as the key controllers of "healthy" aging of elderly people. Immunity & Ageing. 2021;**18**:1-11

[33] Kim S, Jazwinski SM. The gut microbiota and healthy aging: A mini-review. Gerontology. 2018;**64**:513-520

[34] Dupont H. Evidence for the role of gut microbiota in irritable bowel syndrome and its potential influence on therapeutic targets. Alimentary Pharmacology & Therapeutics. 2014;**39**:1033-1042

[35] Morgan M. The treatment of chronic hepatic encephalopathy. Hepato-Gastroenterology. 1991;**38**:377-387

[36] Foster JA, Neufeld K-AM. Gut–brain axis: How the microbiome influences anxiety and depression. Trends in Neurosciences. 2013;**36**:305-312

[37] Mayer EA, Padua D, Tillisch K. Altered brain-gut axis in autism: Comorbidity or causative mechanisms? BioEssays. 2014a;**36**:933-939

[38] Berrill J, Gallacher J, Hood K, Green J, Matthews S, Campbell AK, et al. An observational study of cognitive function in patients with irritable bowel syndrome and inflammatory bowel disease. Neurogastroenterology & Motility. 2013;**25**:918-e704

[39] Koloski NA, Jones M, Kalantar J, Weltman M, Zaguirre J, Talley N. The brain–gut pathway in functional gastrointestinal disorders is bidirectional: A 12-year prospective population-based study. Gut. 2012;**61**:1284-1290

[40] Pimentel M, Lembo A, Chey WD, Zakko S, Ringel Y, Yu J, et al. Rifaximin therapy for patients with irritable bowel syndrome without constipation. New England Journal of Medicine. 2011;**364**:22-32

[41] Crouzet L, Gaultier E, Del'homme C, Cartier C, Delmas E, Dapoigny M, et al. The hypersensitivity to colonic distension of IBS patients can be transferred to rats through their fecal microbiota. Neurogastroenterology & Motility. 2013;**25**:e272-e282

[42] Kennedy PJ, Cryan JF, Dinan TG, Clarke G. Irritable bowel syndrome: A microbiome-gut-brain axis disorder? World Journal of Gastroenterology: WJG. 2014;**20**:14105

[43] Bravo JA, Julio-Pieper M, Forsythe P, Kunze W, Dinan TG, Bienenstock J, et al. Communication between gastrointestinal bacteria and the nervous system. Current Opinion in Pharmacology. 2012;**12**:667-672

[44] Stilling RM, Dinan TG, Cryan JF. Microbial genes, brain & behaviour–epigenetic regulation of the gut–brain axis. Genes, Brain and Behavior. 2014;**13**:69-86

[45] Heijtz RD, Wang S, Anuar F, Qian Y, Björkholm B, Samuelsson A, et al. Normal gut microbiota modulates brain development and behavior. Proceedings of the National Academy of Sciences. 2011;**108**:3047-3052

[46] Clarke G, Grenham S, Scully P, Fitzgerald P, Moloney R, Shanahan F, et al. The microbiome-gut-brain axis during early life regulates the hippocampal serotonergic system in a sex-dependent manner. Molecular Psychiatry. 2013;**18**:666-673

[47] Saulnier DM, Ringel Y, Heyman MB, Foster JA, Bercik P, Shulman RJ, et al. The intestinal microbiome, probiotics and prebiotics in neurogastroenterology. Gut Microbes. 2013;**4**:17-27

[48] Bercik P, Denou E, Collins J, Jackson W, Lu J, Jury J, et al. The intestinal microbiota affect central levels of brain-derived neurotropic factor and behavior in mice. Gastroenterology. 2011;**141**:599-609

[49] Ait-Belgnaoui A, Colom A, Braniste V, Ramalho L, Marrot A, Cartier C, et al. Probiotic gut effect prevents the chronic psychological stress-induced brain activity abnormality in mice. Neurogastroenterology & Motility. 2014;**26**:510 520